Mechanics of Wood Machining

Etele Csanády · Endre Magoss

Mechanics of Wood Machining

Second Edition

Springer

Etele Csanády
Department of Wood Engineering
University of West-Hungary Sopron
ADY E.U. 5
Sopron 9400
Hungary

Endre Magoss
Department of Wood Engineering
University of West-Hungary Sopron
ADY E.U. 5
Sopron 9400
Hungary

First edition published by the Department of Wood Engineering, University of West Hungary, Sopron, 2011

ISBN 978-3-642-29954-4 ISBN 978-3-642-29955-1 (eBook)
DOI 10.1007/978-3-642-29955-1
Springer Heidelberg New York Dordrecht London

Library of Congress Control Number: 2012939641

Printed on acid-free paper

Springer is part of Springer Science+Business Media (www.springer.com)

Foreword

Wood is one of the most precious materials of the mankind. Since ancient times wood material has been widely used and most of the utensils, shelters and dwellings was made of wood. Wood may have exceptional beauty in all shades of colour and many species have excellent durability withstanding many hundreds of years. The oldest wooden buildings have withstood storms and disasters over 1,300 years. At the same time, wood materials require one of the lowest energy inputs on the earth.

Since ancient times the wood has been shaped by mechanical woodworking operations as we do today. We stand with astonishment and true admiration before a 20 m high wooden column, somewhere in China or Japan, worked by hand with incredible accuracy—a thousand years ago!

Today we have modern woodworking machines and hand tools; however, the quality of the raw material is continuously declining. We are forced to use this precious material with care and more economically, reducing waste as much as possible. To fulfil this requirement, efforts are being made worldwide to develop new machines, tools and technologies, for instance thin-kerf saws, wear resistant tools, quality assessment, and using of smaller size raw materials, etc.

The first comprehensive textbook on "Theory of Wood Processing" in Hungarian was published in 1994 and used by students at the Faculty of Wood Science, at the University of Sopron. This textbook contains a special chapter on the "Theory of Wood Cutting" and its related topics. Since the preparation of this book, almost 20 years have passed, and during this time new theoretical and practical knowledge accumulated and some new methods have been developed.

This book is the improved and expanded version of the above mentioned chapter, supplemented with several related topics on vibration and stability of woodworking tools, the phenomenon of washboarding, tensioning methods and pneumatic clamping. The last chapter contains an important topic on wood surface roughness. In the last 15 years, fundamental research work has been done in this field and new approaches were developed. The new methods developed made it possible to characterize the different wood species with numerical values and to find general relationships concerning very different wood species.

It should be stressed that important results and new ideas have been published and, with all probability, will be published in many languages all over the world. Therefore, it is of great importance to consider all significant works done whenever and everywhere.

I hope this new textbook on the "Mechanics of Wood Machining" will contribute to the better understanding of physical phenomena associated with real woodworking processes. The material presented is designed for graduate and postgraduate students, researchers and also to help practising engineers find the pertinent knowledge on this subject.

Sopron, Hungary, November 2011 György Sitkei

Preface

The importance of economical processing and manufacturing of various wood products is everincreasing. During technological processes wood materials may be exposed to various mechanical effects, e.g. the interactions between wood materials and different kind of tools, and the general laws governing these interactions should be known.

The behaviour of wood materials deviates essentially from that of the generally known elastic materials. Wood has a strong anisotropy and viscoelastic properties which have highly aggravated the derivation of generally valid relationships. In the past time, because of our lack of understanding, knowledge of these processes was essentially restricted to empirical relationships and practical recommendations.

This textbook has attempted to develop a mechanics of wood machining in depth and extent so that it defines and delineates an area of knowledge that may be a new discipline in the future. Each topics in this book is developed from fundamental concepts of mechanics. Where possible, mathematical treatment has been developed in quantitative terms to express specific wood-machine relations on the assumption that these relations should obey fundamental laws. Such a mechanics is capable of describing and predicting the interaction of wood and machine in terms of performance, quality and energy consumption. In many cases working diagrams are elaborated to facilitate the consideration of several variables influencing the specific force, energy requirement of surface quality.

This textbook is an attempt to convey both definitive and practical information concerning wood machining to persons interested in the various problems of wood machining. The topics were treated with the view in mind of presenting as much physical information as possible without excessive mathematical manipulation.

A distinct strength of this book is that it draws on truly international literature and not only English language research publications. At the same time, considerable part of this book relies on own research works.

The authors are especially indebted to Professor G. Sitkei for reading the completed manuscript and offering many useful suggestions.

The authors acknowledge the important assistance of Veronika Csanády and Lajos Reisz in the preparation of the manuscript for publication. They are also sincerely grateful to the staff of Springer Verlag for their excellent work done with patience.

March 2012

Etele Csanády
Endre Magoss

Contents

Chapter 1
Mechanics of the Cutting Process

Machining plays a basic role in most procedures of the conversion of timber. Machining usually alters the shape, size and surface quality of wood. Machining occurs by cutting in most cases, and chips are the by-product. Some chips are used for the production of chipboard and some are used to generate energy. The chip may be the product in certain cases, and then we talk about technological or target chips. Chips are deliberately produced for the production of plywood, chipboard and wood-wool for packing.

Woodworking can be divided into two main groups: machining with or without cutting (Fig. 1.1).

The machining without cutting is confined only to specific areas. Plywood cutting and burning designs on surfaces can be done with laser beams. Lasers are not generally used for cutting because they use too much energy have low cutting efficiency. The shape of components can only be changed with bending. That is why an additional cutting operation is usually necessary.

The economic importance of wood, as a major renewable resource, will certainly increase the efforts worldwide to improve tool and machine design and to develop new technologies. For example, tool (saw) vibrations increase the kerf losses, reduce product accuracy, diminish surface quality and reduce tool life. Therefore, increasing attention is being paid to control tool and machine vibration to achieve more efficient cutting. Reducing tool vibration and control can be attained by appropriate tool design, membrane stress modification and process parameters.

Thin kerf sawing is a progressive method to reduce kerf losses. The optimum feed speed a very important process parameter. It is determined by the relationship among gullet capacity, feed speed, saw speed and cutting depth. It is important to avoid over or underfeeding to achieve thin-kerf sawing [1].

In the following chapter we discuss the common woodworking operations and give the necessary relationships and explanations for optimum design and choice of operating parameters.

E. Csanády and E. Magoss, *Mechanics of Wood Machining*,
DOI: 10.1007/978-3-642-29955-1_1, © Springer-Verlag Berlin Heidelberg 2013

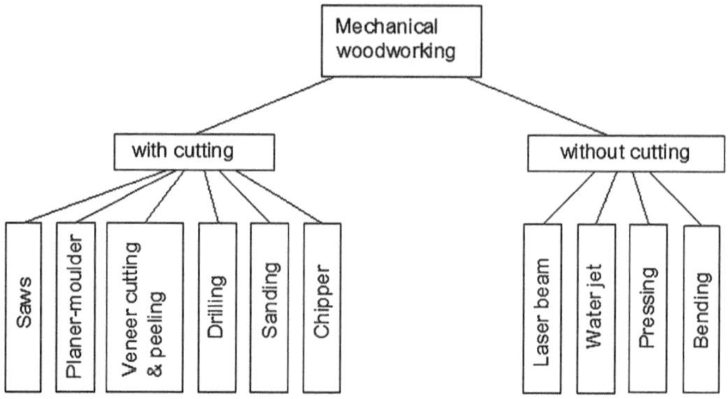

Fig. 1.1 Mechanical woodworking methods

Fig. 1.2 Principal cutting directions and cutting in intermediate directions [2]

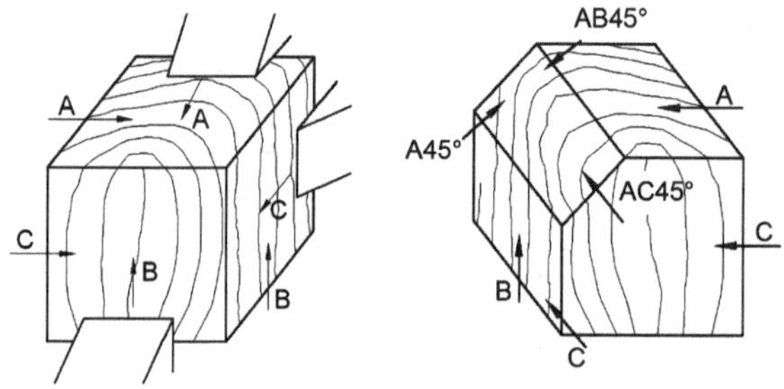

1.1 Definitions and Notations

In machining, cutting results from the interaction of the timber and the tool. The physical and mechanical properties of the timber, the geometry of the tool and the operating parameters determine the interaction. The anisotropy, the compression and the tensile strength, and the elongation at rupture are the most important material characteristics of wood. The main cutting directions corresponding to the anisotropy are demonstrated in Fig. 1.2. When cutting with the grain, the compression strength is about the half of tensile strength, while the tensile strength perpendicular to the fibre is only about one twentieth of the tensile strength along the grain. The elongation at rupture of wood is very small compared to the metal.

The angles and the edges are important characteristics of the tool's geometry (Fig. 1.3) [2].

The following tool angles are important:

α—The clearance angle ensures that there is no friction on the bottom face of the blade. It is generally 10–15°, but veneer cutting knives have only around a 1° clearance angle

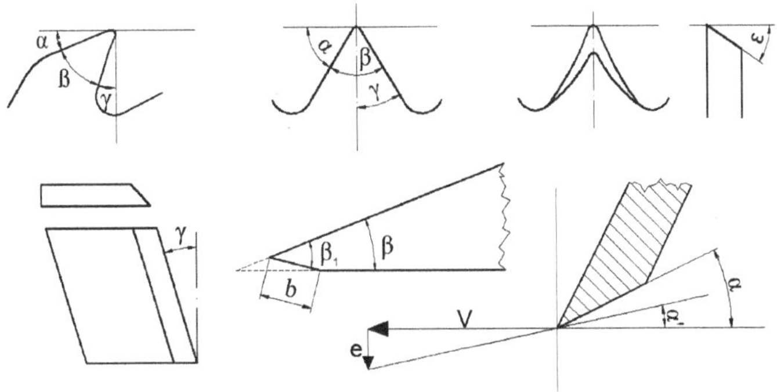

Fig. 1.3 Characteristic angles of woodworking tools [2]

β—The sharpening or bevel angle; its main function is the cutting itself, maintaining a sharp edge and conducting heat away from the edge part. Veneer cutting knives have values around 20°

γ—The rake angle determines the chip deformation and is between 15 and 25°. Very hard tool materials require quite a small rake angle, while veneer cutting tools are around 70°

ε—The top bevel angle generally used on circular saws (alternately or with a double bevelled top) improve sawing accuracy

λ—The oblique or slide angle decreases the normal force on the edge and the dynamic load on long knives. An oblique angle lowers the true cutting angle according to the following equation:

$$\tan \delta' = \tan \delta \cdot \cos \alpha$$

$$\delta = \alpha + \beta$$

where $\delta = \alpha + \beta$ is the cutting angle.

α'—the moving clearance angle occurs in peeling veneer and drilling due to the combine effect of a circular and a linear feed motion:

$$\tan \alpha' = \frac{e}{v}$$

where v is the velocity of circular motion and e is the speed of linear motion. In the presence of a moving clearance angle, the actual clearance angle decreases and can entirely vanish.

β_1—back microbevel angle is used in cutting veneer to avoid excessive wear of the knife edge having a small sharpening angle.

A special form of the microbevel angle is given in jointing. The jointing operation occurs on installed knives where a grinding stone is passed over the knife edges as they turn in the cutter head. The cutting edge itself has a zero clearance angle. The jointed land width is less than 1.0 mm.

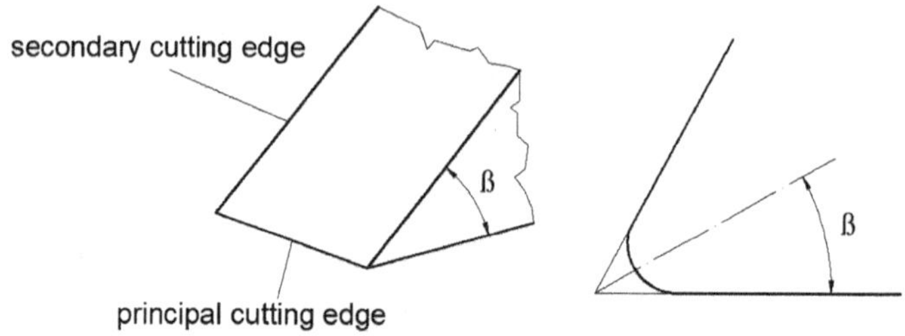

Fig. 1.4 The edges of the tool and the round edge

The edges of the tool are important in the cutting process. We distinguish between principal and secondary cutting edges; the meeting point of the two edges is the *cutting-point*. The *cutting-edges* are never absolutely sharp, but they always have a rounding off radius. The range of the radius of the main-cutting edge is $\rho = 10\text{–}60\ \mu\text{m}$.

If the radius becomes larger due to wear, the tool edge has to be sharpened (Fig. 1.4).

The operational parameters of a tool can be summarized as follows:

- The peripheral speed of the tool (v) that influences its productiveness, the surface quality, the wear of the edges, and to a lesser extent the cutting force. Regular values vary between 30 and 60 m/s today.
- The feed per tooth (e_z), or the chip thickness (h) influence the productiveness, the surface quality and the cutting force. If the gullet of the tooth runs full in a gap, then the chip has to fit loosely in the tooth's gullet, and that limits the maximum value of e_z. We get varying chip cross sections when using rotating tools, so we should calculate the average chip thickness. Using a frame saw, the value of e_z changes as a function of the stroke, as the saw's speed is variable, while the feed motion is constant. When cutting wood, the theoretical thickness of a chip is almost the same as the thickness of the detached chip. The reason is that the chip's plastic deformation is not significant; smaller plastic deformations occur only on the compressed side. When cutting metal, the thickness of the detached chip is two to three times as big as its theoretical thickness.
- The number of edges of a tool (z) influence its productivity, the surface quality, and sometimes the maximum feed per tooth. The pitch of the teeth and the capacity of the tooth gullet are important in saws.

There are relatively simple mathematical relations between the tool parameters and operational factors, which will be given in Chap. 3.

1.2 Mechanics of Cutting

The mechanics of cutting examines the interactions between the tool and the chip using the laws of engineering mechanics [3]. We take into consideration the mechanical properties of the material, if necessary the anisotropy, the viscoelasticity

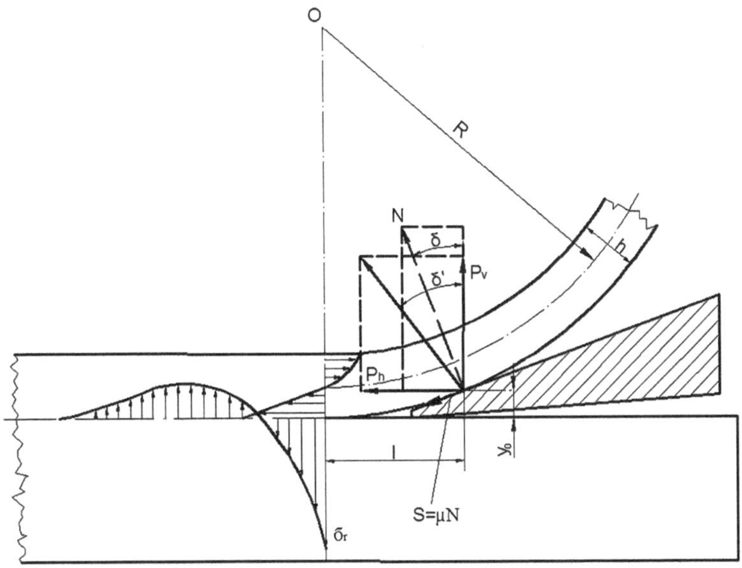

Fig. 1.5 Force and stress relations of the chip [2, 4]

and its plastic properties. We look for relationships between the forces and strains (deformations), and the final goal is the determination of the cutting force depending on the properties of the material and the tool. Our aim is to provide a scientific foundation to the phenomena which so far has only been observed.

We also confirm the theoretical results with experimental data. Where necessary, we give diagrams to more quickly determine the required parameters.

1.2.1 Force and Stress Relations

Certain assumptions usually have to be made at the beginning of theoretical derivations. Our most important initial assumption concerns on the shape of the chip. We suppose that the chip deforms on the knife's surface on a radius *R*. A series of practical observations have confirmed this assumption. e.g. the shavings of a planer come off the tool's front surface in a closely rolled state using cutting direction *B*. The rolled state of the chip remains after cutting (after unloading) because the chip's compressed side has a plastic flow and this means permanent deformation.

The force and stress relations of the chip are demonstrated in Fig. 1.5 [2, 4]. The normal force *N* acts on the tool's front surface, which can be divided into horizontal and vertical components. The chip slides on the front surface; therefore a frictional force $S = \mu N$ arises on the rake face that can be divided into components. The resultant force from the sum of the components shows an inclination towards the direction of the movement. Its angle from the vertical will be given as follows:

$$P_h = N\sin\delta + \mu N\cos\delta$$
$$P_v = N\cos\delta - \mu N\sin\delta$$

from which:

$$\tan\delta' = \frac{P_h}{P_v} = \frac{\tan\delta + \mu}{1 - \mu\cdot\tan\delta} \tag{1.1}$$

where P_h is the horizontal (cutting) force and P_v is the vertical force.

The internal stress distributions are also given in the Figure. The internal stress distribution due to bending is asymmetric because the compressive strength is roughly half of the tensile strength. The compression stress is near constant on the compressed side where the load on the material equals its strength. The result of the asymmetric stress distribution is that the neutral axis shifts to the tension side. Further significant shifting of the neutral axis to the tension side occurs because the horizontal component of the cutting force exerts an axial pressure on the chip, as on a bent beam, so it further increases the width of the compressed zone, which is loaded up to its maximum strength. (This observation is used effectively when bending furniture parts).

The bending of the chip produces stresses in the horizontal level in front of the knife-edge. It is important to know this because in some cases there may be *pre-splitting of the material.* Then the surface is not actually cut by the knife edge, so we get a very rough surface similar to the surface of wood that was split by an axe.

The stress distribution in front of the cutting edge can be calculated according to the following equation:

$$\sigma_{zr} = k\cdot y$$

where:

k the deformation modulus of the wood, N/mm^3;
y the deformation in a radial direction.

The k deformation modulus can be determined approximately with an indenter. In the theory of elasticity the so-called "Boussinesq-problem" gives the following correlation [3]:

$$p = \frac{4E}{\pi(1-v^2)}\cdot\frac{y}{d} = k\cdot y \tag{1.2}$$

where:

d the diameter of the indenter
E the modulus of elasticity of the material in a radial direction
v Poisson's ratio
p pressure

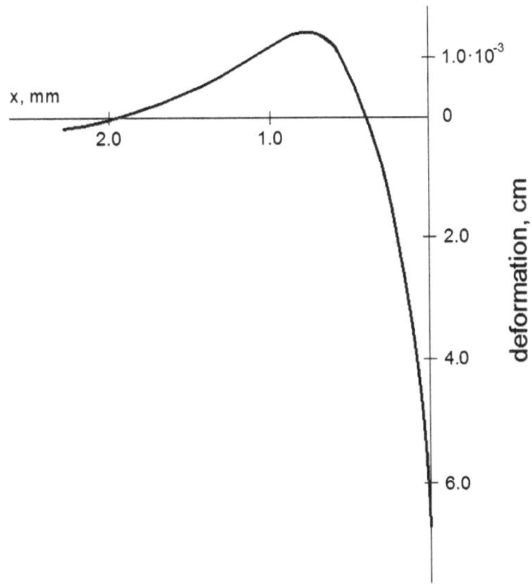

Fig. 1.6 The changing of radial deformation in the plane in front of the edge [2]

The k value depends on the diameter of the indenter according to Eq. (1.2). From our own experiments on oak, using an indenter 6 mm in diameter, we calculated 80–85 N/mm^3.

The y deformation in a distance x from the edge of the knife can be calculated as:

$$y = -\frac{P_v \cdot e^{-\beta \cdot x}}{2\beta^3 E \cdot I}[\cos \beta x + \beta \cdot l(\cos \beta x - \sin \beta x)] \tag{1.3}$$

where:

$$\beta = \sqrt[4]{\frac{k}{4EI}}$$

P_v the vertical force component
I the moment of inertia of the chip
l the distance of the attack point of the P_v force

Using the above equations, the maximum tensile stress occurs under the edge of the knife, its value is given from Eqs. (1.2) and (1.3) as:

$$\sigma_{zr\max} = \frac{kP_v(1 + \beta \cdot l)}{2\beta^3 \cdot E \cdot I} \tag{1.4}$$

We can see the change of radial deformation in a given set of parameters ($h = 0.5$ mm, $\delta = 50°$, $h/R = 0.25$) in Fig. 1.6. The deformation in a given distance x changes from tensile to compressive. Its relative location (x^*) can be seen in Fig. 1.7 depending on the thickness of the chip.

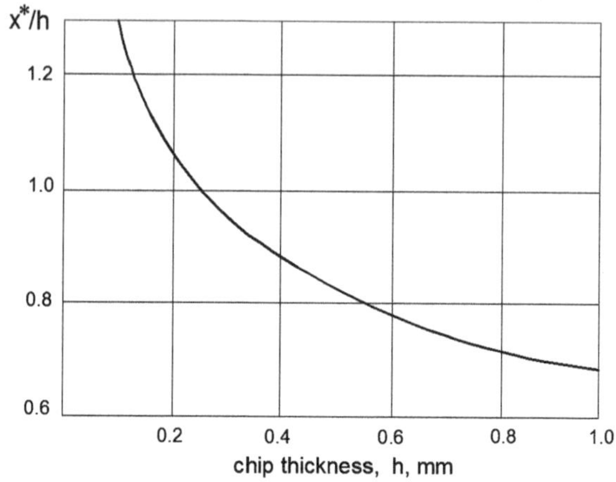

Fig. 1.7 The displacement of the relative zero point in the plane in front of the edge depending on the chip thickness [2]

As the thickness of the chip increases, the relative zero point approaches the edge of the knife, and this implies that the danger of pre-splitting increases as the thickness of the chip increases.

The conditions of pre-splitting can be formulated as follows:

1. The maximum value of the radial load σ_{zrmax} has to be larger than the tensile strength of the wood.
2. The chip may not break, that is that the load on the tensile side of the chip has to be smaller than the tensile strength of the wood.

1.2.2 *The Cutting Force*

In order to determine the components of the cutting force, we use the balance of the moments. If the neutral axis lies in the middle of the beam, then the expressions of the relative elongation and the resistance moment are:

$$\varepsilon = \frac{r - R}{R} = \frac{h}{2R} \quad \text{and} \quad W = \frac{bh^2}{6}$$

These values alter when the neutral axis is dislocated in the following way:

$$\varepsilon \cong \frac{h}{5R} \quad \text{and} \quad W \cong \frac{bh^2}{10}$$

Using the above approximations, the equilibrium moment equation can be given as follows:

$$M = P_v l + P_h \cdot y_0 = \sigma_h \cdot W = E\frac{h}{5R} \cdot \frac{b \cdot h^2}{10} \quad (1.5)$$

where:

b—the width of the chip.

The values of the coordinates l and y_O in the above equation are approximately:

$$l \cong R\sin\delta \quad \text{and} \quad y_0 \cong R(1-\cos\delta)$$

Inserting these into the equilibrium equation we get:

$$P_v = \frac{1}{f(\mu,\delta)} \cdot \frac{E \cdot b}{50} \cdot \left(\frac{h}{R}\right)^2 \cdot h \tag{1.6}$$

where:

$$f(\mu,\delta) = \sin\delta(1-\mu\cdot\tan\delta) + (\sin\delta + x_0/R)\cdot\tan\delta\cdot(\tan\delta+\mu)$$

and the x_O in the last equation means the geometrical pre-splitting; (see Fig. 1.10).

Considering Eq. (1.1), the resultant horizontal force, the cutting force can be calculated from the following equation:

$$P_h = \frac{\tan\delta'}{f(\mu,\delta)} \cdot \frac{E}{50} \cdot \left(\frac{h}{R}\right)^2 \cdot h \tag{1.7}$$

Equation (1.7) gives the force which is spent to the deformation of the chip. In reality the rounded edge is affected by force which has to overcome the compressive strength of the material. This force is proportional to the cross section of the edge and the compressive strength of wood:

$$P'_h = s\cdot b\cdot\sigma_c$$

where:

$s \cong 2\rho$	the thickness of the edge,
ρ	the radius of the edge,
σ_c	the compressive strength of the wood.

The final expression of the cutting force related to the unit length will be as follows:

$$P_h/b = 2\rho\sigma_c + \frac{\tan\delta'}{f(\mu,\delta)} \cdot \frac{E}{50} \cdot \left(\frac{h}{R}\right)^2 \cdot h \tag{1.8}$$

Examining the second term in the above equation, we will come to important conclusions.

The *tan* $\delta'/f(\mu, \delta)$ trigonometric function occurring in the second term of Eq. (1.8) is illustrated in Fig. 1.8. One can see that the diagram has a minimum at about $\delta = 40°$, and that means that the cutting force is minimal in the given range. This statement is totally in accordance with our experimental observations. Using

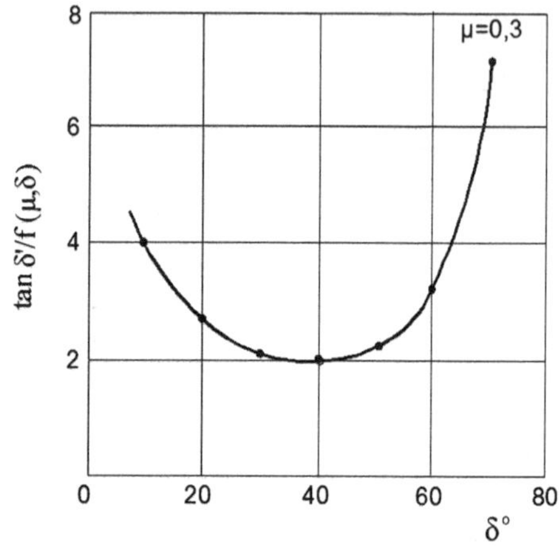

Fig. 1.8 The minimum of the cutting force depending on the δ angle [2, 4]

other friction coefficients only slightly modify the diagram and the location of the minimum.

The values of friction coefficients measured at low surface pressures cannot be used here. The surface pressures are on the order of 100 bar, and here the friction coefficient is smaller. The effect of higher water content is different from the usual one. Water comes to the surface as a result of the pressure, and this lowers the friction coefficient. The high temperatures on the friction surface also considerably decrease the friction coefficient. Therefore we record the values of μ between 0.2 and 0.3.

The Eq. (1.8) is linear depending on the thickness of the chip, so far as the *h/R* ratio is constant at a given δ angle. According to experimental results, Eq. (1.7) is linear with a good approximation in any cutting direction; therefore, it is worthwhile to further study the deformation of the chip.

1.2.3 Chip Deformation

In the previous derivations, we supposed that the point of attack of the cutting force and therefore the *h/R* ratio is known. That is not the case. Chip deformation has more than one degree of freedom and therefore, the *h/R* ratio cannot be calculated from static equilibrium equations. The chip can deform in different ways depending on the combined action of internal and external forces. In order to determine the *h/R* ratio, the variational principle (minimum of potential energy) may be used. The basic idea of this method is to find a chip deformation in which the work done by the forces would be a minimum. To solve this problem, we describe the virtual work produced by internal and external forces as follows

$$\delta(U_i + U_e) = 0 \quad and \quad U_e = -\sum F_j \delta y_j$$

where

U_i, U_e the work done by internal and external forces,
δ symbol of variation,
F_j external forces,
δy_j virtual displacements.

The work done by internal forces in a bent beam is given by the following equation (Ritz method):

$$U_i = \frac{1}{2}\int E \cdot I y''^2 dz = \frac{1}{4}\left(\frac{\pi}{2l}\right)^4 \cdot E \cdot I \cdot y^2 \cdot l$$

where y'' means the curvature of the neutral axis.

The work done by external forces is expressed in the following form:

$$U_e = \frac{P_h \Delta l}{2} + \frac{P_v \cdot y_0}{2}$$

in which the longitudinal deformation of the chip is given by

$$\Delta l = \frac{P_h \cdot l}{E \cdot b \cdot h}$$

Keeping in mind Eq. (1.1), the tangential and radial force components can be expressed using the bent beam theory:

$$P_h = \frac{3EIy_0}{l^3}(\tan\delta + \mu)$$

$$P_v = \frac{3EIy_0}{l^3}(1 - \mu\tan\delta)$$

Taking the functional $F = U_i - U_e$ and its derivative with respect to y, setting the functional equal to zero, we get

$$\frac{\partial F}{\partial y} = 0$$

Solving the above equation and using the $l \cong R \sin\delta$ approximation, the h/R ratio is given by the following equation:

$$\frac{h}{R} = \frac{2\sin\delta}{(\tan\delta + \mu)}\sqrt{\mu\tan\delta} \tag{1.9}$$

In the above derivation we have supposed that the total bending work occurs in the elastic range and no plastic deformation occurs, but that is not the case. Beyond the yield limit, the bending forces remain constant and do not increase linearly with the deformation. Closer examination of this distortion gave the following correction:

$$\frac{h}{R} = \frac{2\sin\delta}{\sqrt{3}(\tan\delta + \mu)}\sqrt{\mu\tan\delta} \tag{1.9a}$$

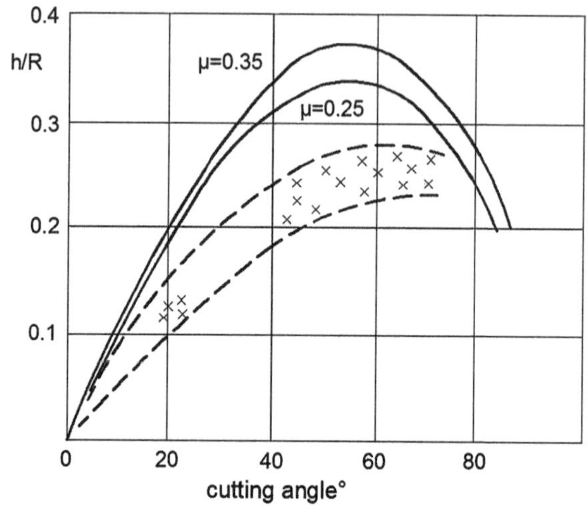

Fig. 1.9 *h/R* ratio as a function of cutting angle for two friction coefficients. (*Dotted line* is the boundary for measurement data) [2]

We can see from the equation that the *h/R* ratio depends only on δ and μ, and these are constant for a given tool. It follows that Eq. (1.8) is linear depending on the *h* chip thickness. In simpler terms:

$$P_h/b = A + B \cdot h \tag{1.8a}$$

Figure 1.9 graphically demonstrates Eq. (1.9), where we have marked measurement results. The values of the measuring results are systematically smaller than those obtained theoretically. This can be explaining by the fact that the upper side of the chip yields plastically, and this leads to the rotation of the bent cross section, thus reducing the apparent value of the δ angle (Fig. 1.10). Therefore the examination of the chip deformation has to be expanded in the plastic state.

1.2.4 Chip Deformation with Plastic Flow

Due to the high load, the upper, compressed side of the chip practically always yields. (The load reaches and surpasses the strength of the chip). As a consequence, the cross section at the foot of the chip turns in a forward direction, and the δ angle apparently decreases (δ^*). Figure 1.10 represents the rotation of the cross section and the decreasing of the geometric pre-splitting (x_0). *Geometric pre-splitting* begins because the plane surface of the tool and the arc-shaped chip touch at one point. We can see from the Figure that the value of x_0 decreases quickly as the δ^* angle decreases. This decreased value of x_0 can decrease further, and can cease completely so that the convex surface of the chip locally deforms, and the theoretical contact line turns into a contact surface.

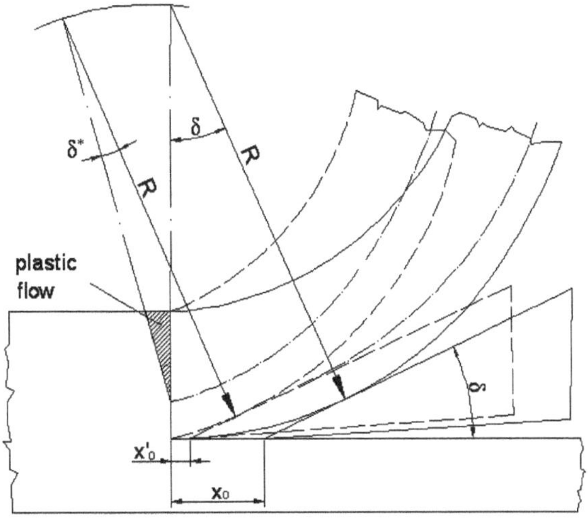

Fig. 1.10 Plastic chip deformation [2, 4]

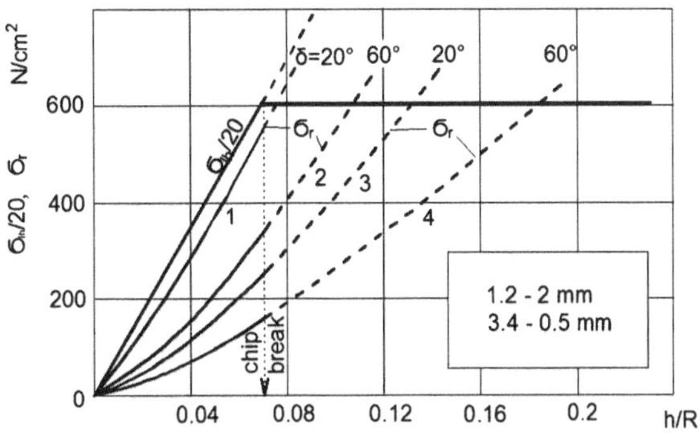

Fig. 1.11 Chip tensile stress due to bending (σ_{lh}) and radial tensile stress in the separation zone (σ_{r}) as a function of the *h/R* ratio for two different chip thicknesses. (To the explanation of the pre-splitting phenomenon before the knife edge) [2]

The experimental values shown in Fig. 1.9 can be explained as follows. Due to the rotation of the cross section, the range of $\delta = 40\text{–}60^\circ$ turns into $\delta^* = 15\text{–}25^\circ$. If we read the *h/R* ratio of the lower values from the Figure, then they almost match the measured values.

In following we have to determine the geometric conditions of pre-splitting. In Fig. 1.11, the chip tensile stress due to bending $\sigma_{\mathrm{lh}}$and the radial tensile stress in the separation zone σ_r are demonstrated as a function of the *h/R* ratio for two different chip thicknesses.

Because tensile strength of wood in the direction of the grain is approximately 20 times larger than the radial tensile strength, we have reduced it by 20 $\left(\frac{\sigma_{lh}}{20}\right)$. The radial tensile stress depends on the cutting angle and chip thickness, so we have constructed several σ_r curves to demonstrate the effect of these variables. Pre-splitting will obviously happen, if the σ_r curve for a given case reaches the horizontal strength boundary line before the chip would break. As we can see from the Figure, this will generally be realized only for chips that are $h \geq 2\,\mathrm{mm}$ thick.

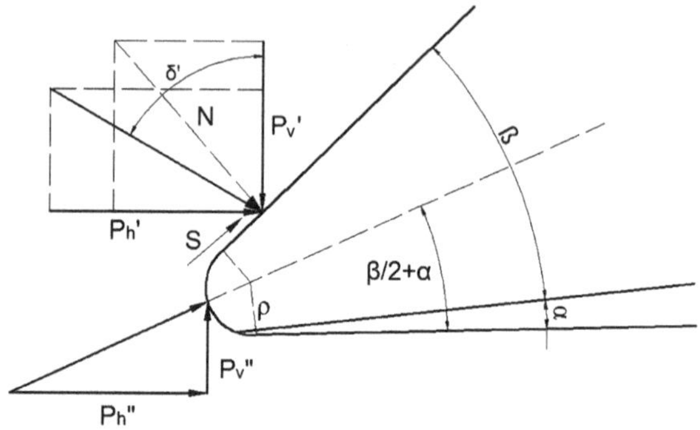

Fig. 1.12 The position of the force acting on the edge [2, 4]

The use of smaller cutting angles is more inclined to cause pre-splitting. For making veneers, we use a cutting angle around 20°. Therefore, the use of a pressure bar is always essential to hamper pre-splitting.

In the practice we rarely produce chips thicker than 2 mm. Therefore, pre-splitting seldom happens.

1.2.5 The Radial Cutting Force

In the derivation of Eq. (1.8) we have supposed for simplicity that the resultant force on the edge is horizontal, (parallel with the direction of cutting). In reality the stress distribution on the edge land is not symmetric, so the resultant force closes up an angle with the direction of motion. The reason is that the extension of the material under the edge of the tool is relatively big, while only an h thickness material is above the edge, (the open wall problem).

The direction of the resultant force is not known exactly and it also depends on the thickness of the chip. Therefore we take, as a first approximation that the direction of the resultant force falls into the bisector of the edge of the knife (Fig. 1.12).

The resultant force, acting on the edge is [see in Eq. (1.8)]:

$$P_e \cong 2\rho b\sigma_c$$

and its components can be written as:

$$P_{eh} = P_e \cos(\beta/2 + \alpha) \quad \textit{and} \quad P_{ev} = P_e \sin(\beta/2 + \alpha)$$

We can write the vertical component of the whole resultant force acting on the tool as follows:

$$P_v/b = 2\rho\sigma_c \sin(\beta/2 + \alpha) - \frac{B \cdot h}{\tan \delta'} \tag{1.10}$$

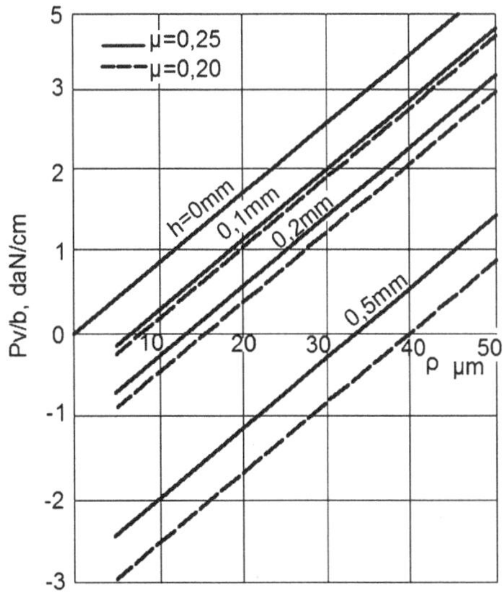

Fig. 1.13 The vertical force component depending on the rounded off radius of the edge [2, 4]

The horizontal component has the form:

$$P_h/b = 2\rho\sigma_c[\cos(\beta/2+\alpha) + \mu \cdot \sin(\beta/2+\alpha)] + B \cdot h \tag{1.11}$$

where:

$$B = \frac{\tan\delta'}{f(\mu,\delta)} \cdot \frac{E}{50} \cdot \left(\frac{h}{R}\right)^2$$

The numerical evaluation of Eq. (1.10) is shown in Fig. 1.13. The vertical component changes its sign depending on the thickness of the chip and the radius of the edge. The force component is practically always negative (tension) when chips are large.

The ratio of the two components can be determined from the derived equations. We can see the P_v/P_h ratio as a function of edge radius in Fig. 1.14. (In both Figures $\beta = 60°$, $\alpha = 15°$ and $\sigma_c = 600$ daN/cm^2.) We also see the change of the sign.

It is known from measurements that by using rotating tools, the radial force component as a function of angular motion has a different shape depending if the tool is sharp or blunt (Fig. 1.15). This phenomenon can now be explained clearly in Fig. 1.13. The thickness of the chip always alters when using rotating tools. If the tool is sharp, then chips 0.2–0.3 mm thick produce a negative radial force and the force turns into a positive value as the chip thickness decreases. Blunt tools almost always produce a positive radial force.

The radial force is always counteracted by the guiding and bearing of the tool. Large positive radial forces (from blunt tools) load the bearing and decreasing its lifetime, but they also have other effects. The positive radial force is an exciting

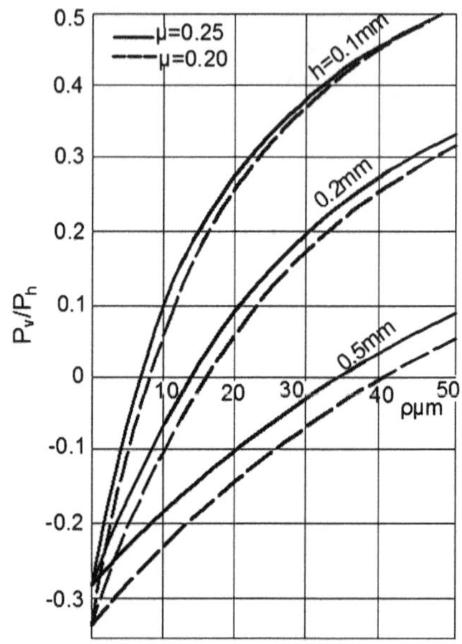

Fig. 1.14 The ratio of the vertical and horizontal force components [2, 4]

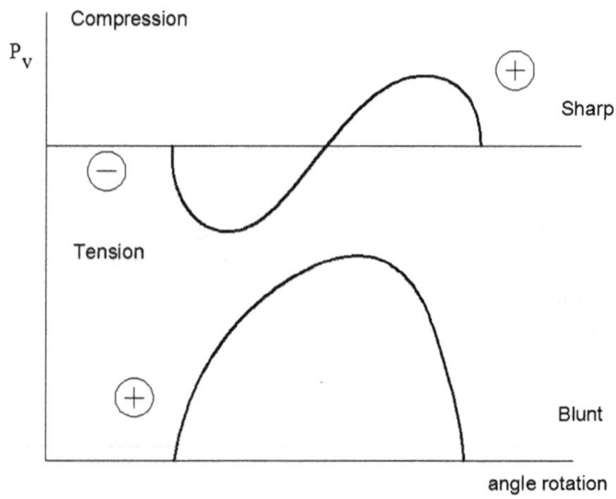

Fig. 1.15 The course of the vertical force for sharp and blunt tools depending on the angle of rotation [2]

force that induces the vibration of the work-piece and the rotating tool. The vibrations worsen the quality of the surface.

In practice, radial forces increase as a consequence of edge wear and jointing. Jointing is a common practice to produce the same cutting circle for all knives of a cutterhead or a peripheral knife planer. Jointing allows several re-sharpening without removing the knives from the cutterhead. The jointed land at the cutting edge has a zero degree clearance angle which can compress a thin layer of the planed board causing cell damage and heat generation due to excessive friction.

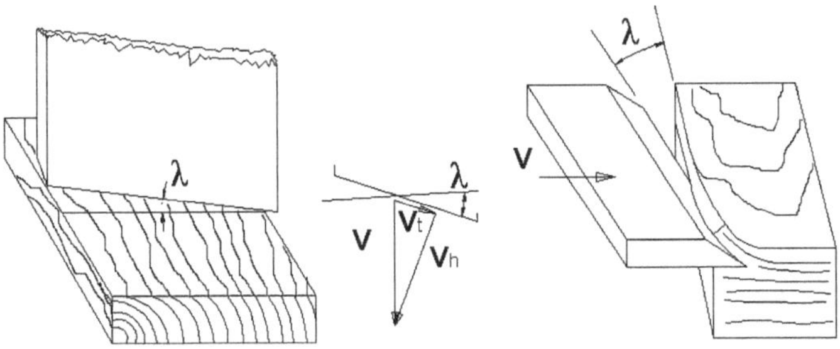

Fig. 1.16 Kinematic relations of oblique cutting and penetration of the knife into the material [6]

Measurements show that the jointing increases the radial force component as jointing land width increases. (Jointed land width is generally less than 1.0 mm).

1.3 Mechanics of Oblique Cutting

The oblique cutting of materials in the metal and wood industry and also in agriculture has been widely known and used for long time [5].

The main advantages of oblique cutting can be summarized as follows:

- the cutting force can considerably be reduced. If the possible counterbalance force of the material is low compared to the cutting force, then the slide cutting method can only be used on soft plastic and upholstering materials.
- the reduction of normal force always means a smaller dynamic load in the driving and bearing mechanisms and also a smaller excitation force generating workpiece vibration [2].
- one of the main advantage of oblique cutting is lowering the noise level, which may be as high as 6–8 dB.
- it may also be assumed that oblique cutting produces a better surface quality.

It seems to be advisable to express the force reduction function in the following dimensionless form [5]:

$$(P_0 - P_\lambda)/P_0 = f\{(\delta - \delta')/\delta\} \tag{1.12}$$

where P_λ is the force at a given oblique angle λ and δ' are the effective cutting angle at a given λ. The cutting energy also varies as a function of oblique angle and it shows a minimum of about $\lambda = 20$–25° [6].

The kinematic relation of the oblique cutting is given in Fig. 1.16. The cutting edge is not orthogonal in the direction of motion but it has an oblique angle (slide angle) λ and the resultant speed consists of two components:

$$\tan\lambda = \frac{v_t}{v_n} \quad \text{and} \quad v_t = v \cdot \sin\lambda, \quad v_n = v \cdot \cos\lambda, \tag{1.13}$$

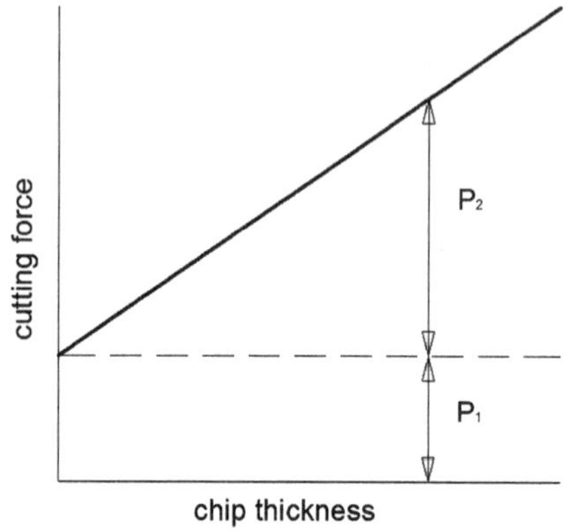

Fig. 1.17 Cutting force versus chip thickness [2]

The value of tan λ is often termed the sliding coefficient. The effective cutting angle δ' varies as a function of the oblique angle according to the following equation:

$$\tan \delta' = \tan \delta \cdot \cos \lambda \tag{1.14}$$

With increasing slide angle, the effective cutting angle can considerably be reduced.

The cutting force consists of two components: one acts on the round edge surface and the second one is due to chip deformation [see Eqs. (1.8) and (1.8a)]. The ratio of these components varies with chip thickness (Fig. 1.17).

The force component acting on the edge depends on the edge radius (dullness) and on the compressive strength of the material. The force due to chip deformation depends on the cutting angle and chip thickness. One of the most characteristic features of oblique cutting is the presence of shear stresses in the separation zone. It is well known that shear stresses contribute to the failure of materials and the equivalent stress is given by the relation:

$$\sigma_e^2 = \sigma^2 + (2 \cdot \tau)^2 \tag{1.15}$$

In cutting, the shear stress is proportional to the normal stress and to the friction coefficient, that is $\tau = \mu\sigma$. The work done by the shear deformation is proportional to the tangential speed component. Therefore, the friction power on a given surface F is expressed by the relation:

$$W_{fr} = \mu\sigma \cdot F \cdot v \cdot \sin \lambda \tag{1.16}$$

Keeping in mind Eq. (1.15), the force acting on the edge can be expressed as a function of oblique angle in the following manner:

$$\begin{aligned} &\sigma_\lambda \cdot F = \sigma \cdot F - 2\mu\sigma \cdot F \cdot \sin \lambda \\ &or \\ &P_\lambda = P_0(1 - 2\mu \cdot \sin \lambda) \end{aligned} \tag{1.17}$$

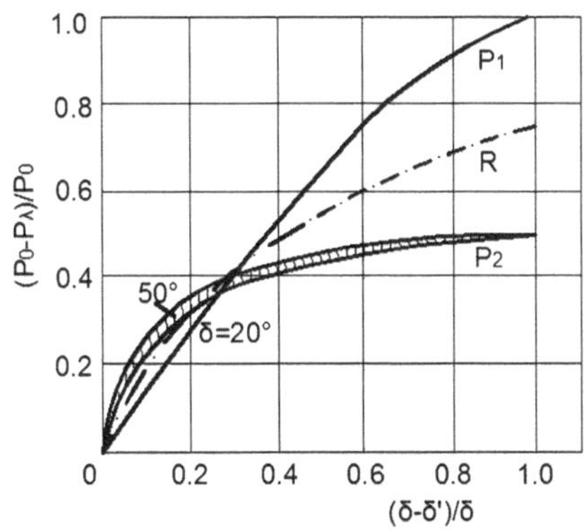

Fig. 1.18 Theoretical force reduction as a function of reducing the cutting angle [6]

Using Eq. (1.17), the relative force reduction for the first force component will be

$$\frac{(P_0 - P_\lambda)}{P_0} = 2\mu \cdot \sin\lambda \tag{1.18}$$

The second force component (P_2 in Fig. 1.17) is due to chip deformation. The following relations are known from the theory of wood cutting [2, 4]: the force is proportional to the square of the relative chip deformation *h/R* and *h/R* is proportional to the 0.75th power of the cutting angle, that is:

$$P \sim (h/R)^2 \quad \text{and} \quad h/R \sim \delta^{0,75} \quad \text{therefore} \quad P \sim \delta^{1,5}$$

Keeping in mind the above discussion, the relative force reduction of the second component is:

$$\frac{(P_0 - P_\lambda)}{P_0} = \frac{\left(\delta^{1,5} - \delta'^{1,5}\right)}{\delta^{1,5}} \tag{1.19}$$

Equations (1.18) and (1.19) are plotted in Fig. 1.8. If we assume for simplicity that both components have the same magnitude, then the resultant force reduction will be given by the geometric mean calculated from the components. This line (*R*) is also plotted in Fig. 1.18. These considerations show that force reduction is not uniquely determined by reducing the cutting angle. To a certain extent, the force reduction will also be influenced by the cutting angle (P_1 component) and by the P_1/P_2 ratio (resultant force).

The penetration of the edge into the material varies considerably if oblique cutting is used (Fig. 1.19). As a consequence of the inclined position, the chip will be lifted from one side before cutting takes place and, therefore, the cutting occurs under the combined action of tensile, compressive and shear stresses. The stretching tensile stress acting vertically to the shear plane reduces the compressive stress under the edge during cutting and produces a better surface quality.

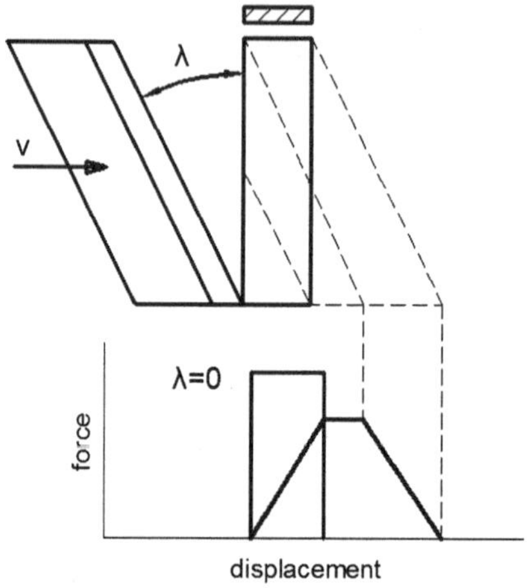

Fig. 1.19 Force displacement relation for different slide angles [6]

Oblique cutting transforms the rounded edge into the section of an ellipse having a smaller curvature radius than that of the edge $\rho' = \rho \cos\lambda$.

A further advantage of the oblique cutting is a smaller rate of increase of force at the start of cutting (Fig. 1.19). The amount of force increase for rotating tools is given by the following equation:

$$\frac{dP}{dt} = \frac{P_\lambda \cdot R \cdot \omega}{\tan \lambda} \tag{1.20}$$

where

R is the tool radius,

ω is the angular speed.

The amount of force rise will influence the dynamic load in driving and bearing mechanisms and also the noise level generated by the tool.

Figure 1.19 shows that the displacement of the knife in the material varies as different oblique angles are used. Since the work done by the edge also depends on its displacement, the cutting energy can be smaller or greater compared to orthogonal cutting.

The work done by orthogonal cutting ($\lambda = 0$) is expressed as:

$$A_0 = P_0 \cdot L \cdot b$$

In oblique cutting

$$A_\lambda = P_\lambda \frac{b^2}{\sin \lambda} + P_\lambda \frac{b}{\sin \lambda}(L \cdot \tan \lambda - b) = \frac{P_\lambda \cdot L \cdot b}{\cos \lambda}$$

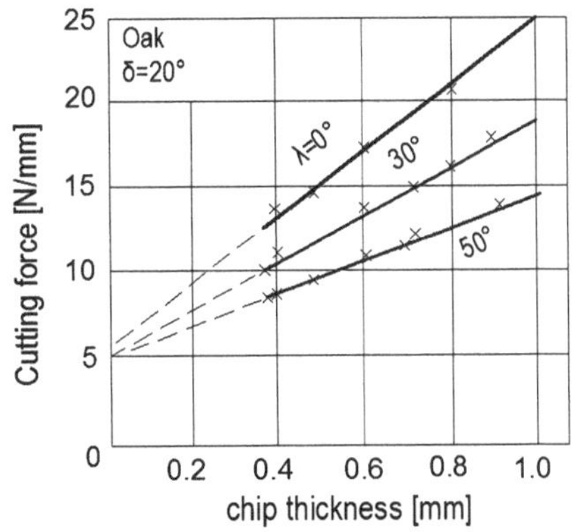

Fig. 1.20 Cutting force versus chip thickness for different oblique angles [6]

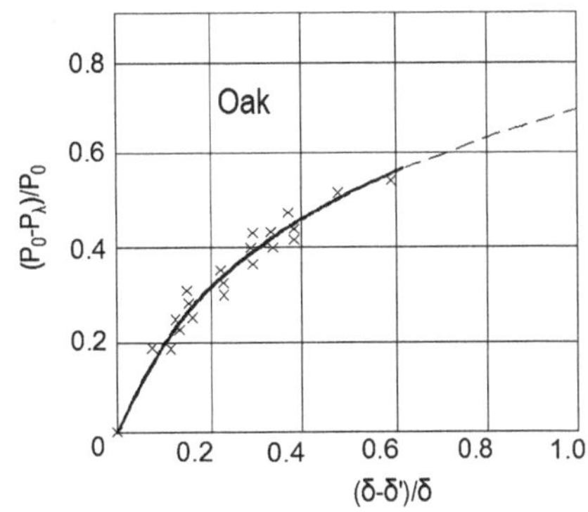

Fig. 1.21 Dimensionless relationship describing force reduction [6]

The ratio of the work performed by these two cutting methods is expressed by

$$\frac{A_\lambda}{A_0} = \frac{P_\lambda}{(P_0 \cos \lambda)} \tag{1.21}$$

where

L is the length of cutting,

b is the width of cutting.

Measurement series for a given oblique angle and cutting angle will be demonstrated as a function of chip thickness. Figure 1.20 shows the results in veneer cutting range. The specific cutting force is linearly related to the chip thickness as predicted theoretically [2, 4]. The curves for different oblique angles show a definite influence of the sliding motion on cutting forces.

The specific force reduction was calculated from the measured cutting forces and the functional relationship was established according to Eq. (1.12). This experimental relationship is plotted in Fig. 1.21. The curve obtained experimentally is surprisingly similar to Fig. 1.18, which is worked out theoretically. The

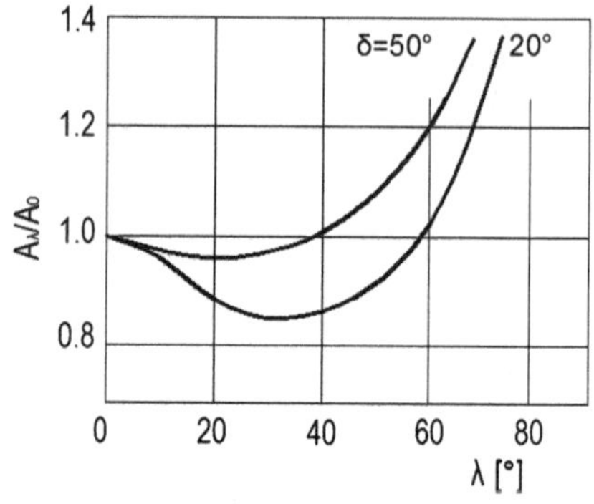

Fig. 1.22 Cutting energy versus oblique angle at two cutting angles [6]

slight scattering of experimental results is explained by the reasons mentioned earlier in Fig. 1.18.

The general relationship given in Fig. 1.21 is expressed by the following half-empirical equation:

$$\frac{(P_0 - P_\lambda)}{P_0} = 0,75\left\{\frac{(\delta - \delta')}{\delta}\right\}^{0,6}$$

The relative cutting energy can be calculated by using the measured cutting forces and Eq. (1.21). These results are shown in Fig. 1.22. The optimum oblique angle from an energetic point of view is a function of the cutting angle and for common cutting angles the same values (20–25°) are valid as established earlier for farm machinery [5]. In the veneer cutting range, however, optimum values are obtained at about $\lambda = 35°$. Another important finding is that considerable reduction in cutting energy can only be achieved with small cutting angles.

Using large oblique angles, the cutting energy always increases due to excess friction.

The surface roughness of the specimens varied in a wide range (R_z values between 40 and 90 μm) and, therefore, a method which shows more information was used to calculate the relative reduction in surface roughness for different oblique angles. Because of scattering of experimental data, only average reduction values may be given. The average relative reduction and its scattering zone are plotted in Fig. 1.23. In practice, roughness can be reduced 15–20 %.

1.4 Mechanics of Bundle Clipper

Clipping a veneer bundle is a special case in woodworking. The bundle is generally 4–6 cm high. The bundle is pressed down with a nose bar before the start of the cutting. Figure 1.24 shows the force relations in the cutting material. The force is composed of two main parts:

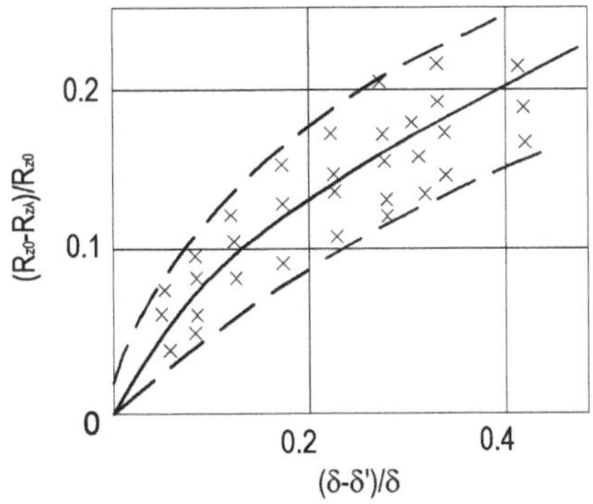

Fig. 1.23 Relative surface roughness reduction as a function of reducing the relative cutting angle [6]

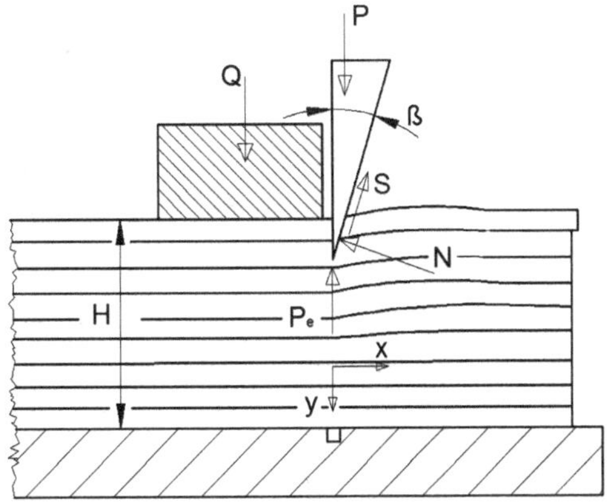

Fig. 1.24 The force-relations of the bundle clipper

- the force acting on the edge and
- the force from deformation and shifting of the material.

The normal force N acting on the face of the knife and the frictional force $S = \mu N$ can be divided into components which can be summarized:

$$P_h = N(\cos\beta - \mu\sin\beta)$$
$$P_v = N(\sin\beta + \mu\cos\beta)$$

The normal force N can be calculated approximately in the following way. The elemental vertical force related to the unit width is:

$$dP_v = E\varepsilon dx$$

and keeping in mind that

$$dx = \tan\beta dy \quad and \quad \varepsilon = \frac{h}{H}$$

where h is the deformation within one layer, we get

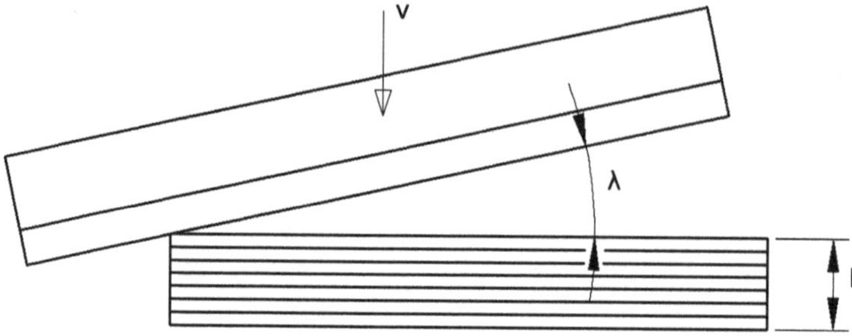

Fig. 1.25 Bundle clipper knife with its slide angle

$$P_v = \left(\frac{E\tan\beta}{H}\right)\int_0^{h_i} hdy = \frac{Eh_i^2}{2H}\tan\beta \tag{1.22}$$

Knowing the vertical force component due to deformation, the normal force N on the face of the knife will be

$$N = \frac{Eh_i^2}{2H\cos\beta}$$

Due to the sliding motion, the normal force N generates a friction force and the total vertical force due to deformation has the final form

$$\sum P_v = \frac{Eh_i^2}{2H}(\tan\beta + \mu)$$

The force component acting on the edge can be calculated as

$$P_e = 2\rho b\sigma_c$$

where ρ is the edge radius, b is the width (length) of the knife and σ_c is compressive strength of the wood.

The total vertical force on the knife is the sum of these two force components which yields

$$P_e = 2\rho b\sigma_c + \frac{Eh_i^2}{2H}(\tan\beta + \mu) \tag{1.23}$$

When the upper layer is cut, this layer will be shifted aside and it exerts no resistance to the knife.

The bundle clipper also has a sliding angle λ (oblique cutting, Fig. 1.25) and the length of the cut is $L = H/sin\lambda$.

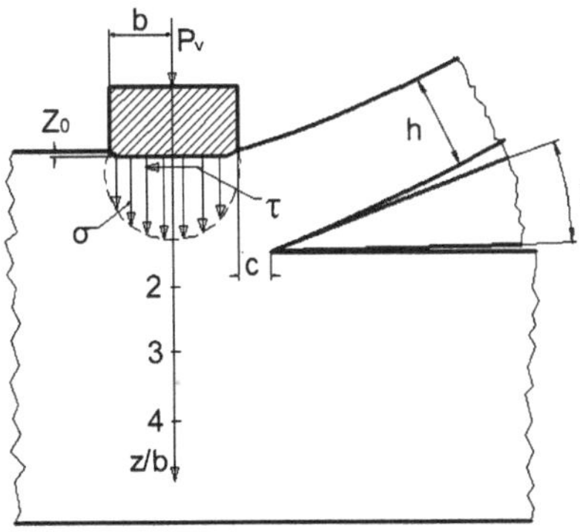

Fig. 1.26 The mechanics of a pressure bar [2]

1.5 Mechanics of the Pressure Bar

Veneer is generally produced by plane cutting or peeling. In order to produce high quality veneer, the peeling machine, especially its knife and nose bar, should always be set in its optimum position. Furthermore, the wood material will be cut when wet and heated ensuring more plasticity to avoid surface checking (cracking).

The main factor governing surface checking is the radius of deformation of the veneer on the rake face of the knife. Therefore, the cutting angle of the knife ($\delta = \alpha + \beta$) is as small as possible, generally 20°. The clearance angle varies between 0.75 and 1.5°. The small bevel angle makes the edge very sensitive to wood defects. In some cases, an attempt was made to decrease the tool edge sensitivity by using a microbevel angle of 25–30° (see in Fig. 1.3 and later in Sect. 3.6).

An important measure in quality veneer production is the use of a pressure bar (nose bar) setting over the knife in an appropriate position (Fig. 1.26). The deformation and the position of the nose bar are related to the veneer thickness in the following way:

The relative deformation is z_0/h; the horizontal distance between the rear edge of the nose bar and the cutting edge related to the veneer thickness c/h is the *relative lead* (see Fig. 1.26).

The pressure exerts a vertical stress field in the wood lying under the bar. This vertical stress has beneficial effects in different ways: it diminishes the splitting of the wood in the plane in front of the cutting edge ensuring a pure cut surface by the tool edge. Furthermore, it may also be assumed that under compression stresses the cutting process is more stable producing smoother surface.

The pressure bar contacting the wood surface may have many different shapes. Some possible shapes are given in Fig. 1.27. Because the pressure bar slides in one direction, it should have rounded edges or a triangle cross section. In order to exert a given vertical stress, the pressure bar should have either a prescribed deformation on the wood surface or a given load per unit length (N/cm).

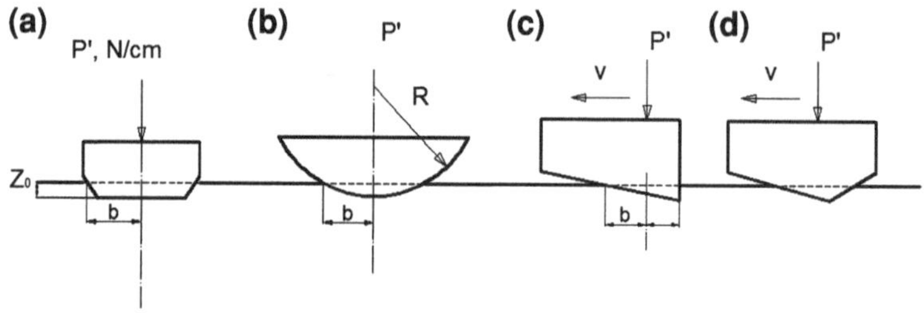

Fig. 1.27 Possible contact shapes for pressure bars

Fig. 1.28 Pressure distribution on the centre line under an infinitely long strip load. *1* triangular pressure distribution; *2* parabolic distribution; *3* constant pressure [3]

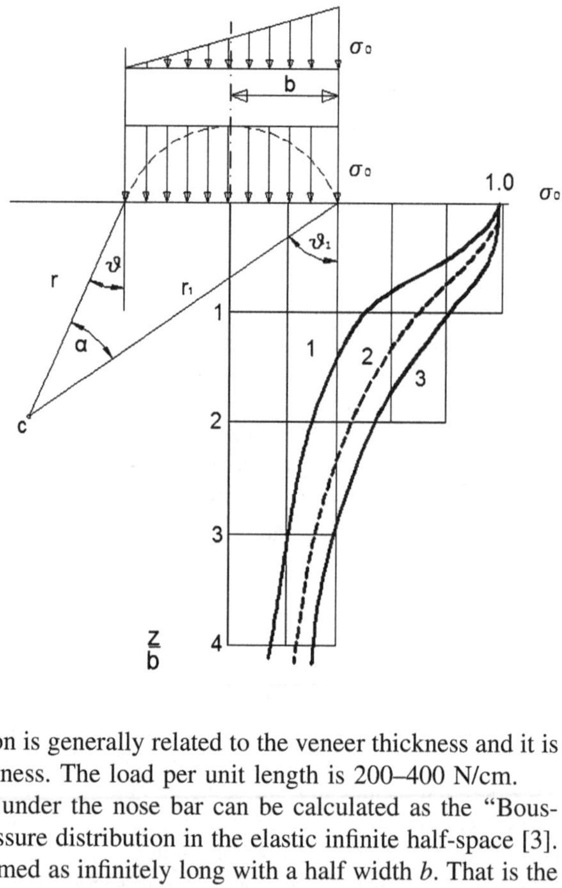

The prescribed deformation is generally related to the veneer thickness and it is 10–15 % of the veneer thickness. The load per unit length is 200–400 N/cm.

The pressure distribution under the nose bar can be calculated as the "Boussinesq problem", i.e. the pressure distribution in the elastic infinite half-space [3]. The contact area can be assumed as infinitely long with a half width *b*. That is the case of a strip load with different pressure distributions on the contact surface, as shown in Fig. 1.28. The Figure shows three different pressure distributions on the contact area and the pressure distribution on the centre line under the contact area as a function of relative depth where stress is exerted (*z/b*).

For a uniform strip load, the vertical stress component at a given point *C* is given by the equation (see Fig. 1.28):

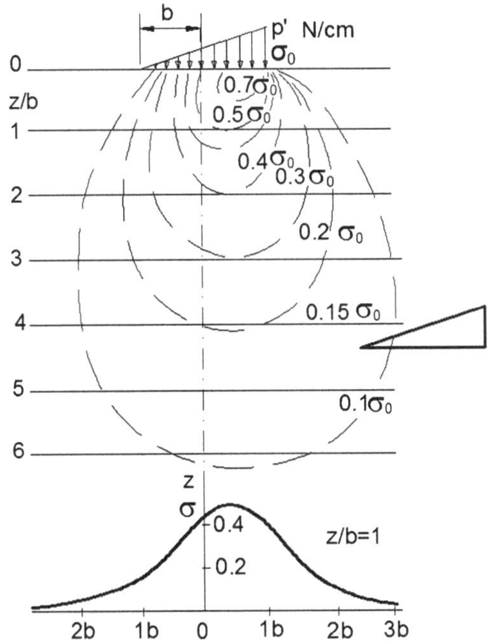

Fig. 1.29 Pressure distribution under a nose bar with triangular loading shape and pressure distribution in a plane located at $z/b = 1$

$$\sigma_z = \frac{\sigma_0}{\pi}[(\vartheta - \vartheta_1) + \sin\alpha] = \frac{\sigma_0}{\pi}(\alpha + \sin\alpha)$$

On the centre line, the angle α can be calculated as

$$\frac{\alpha}{2} = arctg\left(\frac{b}{z}\right)$$

and therefore

$$\sigma_z = \frac{\sigma_0}{\pi}\left[2arctg\left(\frac{b}{z}\right) + \sin\alpha\right] \tag{1.24}$$

which is represented graphically in Fig. 1.28. It is important to note that the vertical stress quickly decreases as a function of depth and the relative depth z/b must always be used.

When a cylinder contacts a plane surface the distribution of pressure along the width of the contact surface is represented by a semi-ellipse. The average contact pressure is:

$$\sigma_0 = \frac{2P'}{\pi \cdot b}$$

where P' is the load per unit length and the centre line vertical stress decreases more rapidly compared to a uniform distribution.

Fig. 1.30 Distribution of shear stresses under a flat nose bar for different friction coefficients [7]

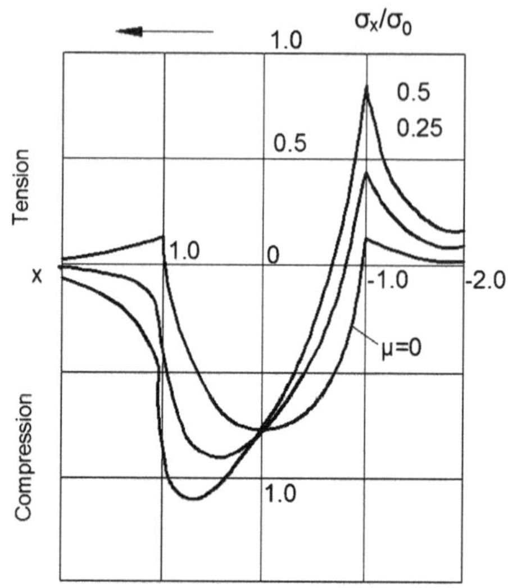

With a triangle pressure distribution in the contact area, we obtain an asymmetric stress distribution in relation to the centre line (Fig. 1.29). In the Figure we can see the vertical stress distribution in a given ($z/b = 1$) relative depth and a possible position of the peeling knife. In this particular case only some 12 % of the maximum contact pressure is available at the edge of the knife. If we would like to have a greater vertical pressure on the knife, a wider contact area must be chosen which means a smaller relative depth for the same veneer thickness. A wider contact area will simply be achieved by using a greater line load P' or deformation $z_{0,}$ as shown in Fig. 1.27.

If we consider a pressure bar profile shown in Fig. 1.27c with an inclination angle of 15°, the following approximations are valid:

$$P' = 4b\sigma_0 \quad \text{or} \quad P' \cong 0,4665\frac{Ez_0}{(1-\upsilon^2)}$$

and

$$\sigma_0 \cong 0,1166\frac{E}{1-\upsilon^2}\cdot\frac{z_0}{b}$$

where E is the modulus of elasticity in the direction of normal pressure and υ is Poisson's ratio. Selection of the appropriate modulus of elasticity may give some trouble. The pressure bar slides on a cut surface with roughness and the small deformation z_0 comparable to the surface roughness. Therefore, the actual value of E may be much less than the measured value on a specimen.

The pressure bar slides on the wood surface and, depending on the friction coefficient, shear stresses develop in the contact area. This generates a force component in the direction of motion $P_h{}' = \mu\, P'$. The total peeling force will be the sum of the cutting force and the friction force.

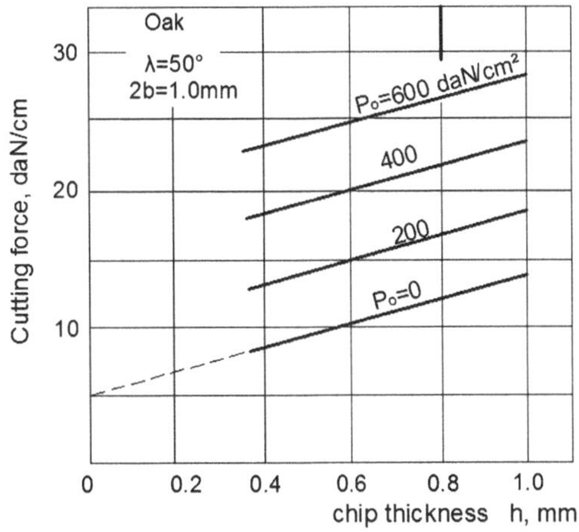

Fig. 1.31 Cutting force using a nose bar with different normal pressures [2]

A log is peeled in tangential direction perpendicular to the grain. Wood has little tensile strength in this direction and therefore, small tensile stresses in this plane can cause checks or cracks. At the front edge of the nose bar the wood surface suffers compression stresses, but at the rear of the contact tensile stresses occur (Fig. 1.30) [7]. This tensile stress depends on the friction coefficient and on the maximum normal stress σ_0. Besides, it is wise to make the rear edge rounded to decrease the peak of the tensile stress.

Thin veneer will be produced by using oblique cutting. As explained in 1.3, cutting with a slide angle λ reduces the normal stresses on the cutting edge and the shear stresses contribute to the cutting process. In this case the total peeling force can be approximated as follows:

$$P_h = A + B_0 \cos\lambda \cdot h + \mu 2b\sigma_{av} \tag{1.25}$$

where A and B_0 are the constants in the cutting force equation for $\lambda = 0$ and σ_{av} is the average pressure on the nose bar. The μ friction coefficient has values between 0.32 and 0.38.

Similar test results for oak wood are seen in Fig. 1.31 using an oblique angle $\lambda = 50°$ and a flat nose bar with rounded edges having the half width $b = 0.5$ mm.

References

1. Kirbach, E.: Saw performance critical to lumber quality, yield. Wood Technol. **39**, 22–24 (1995)
2. Sitkei, G., et. al.: Theorie des spanens von Holz. Fortschrittbericht No. 1. Acta Fac. Ligniensis Sopron (1990)
3. Timoshenko, S., Goodier, J.: Theory of Elasticity. Mc. Graw-Hill, New York (1951)
4. Sitkei, G.: Fortschritte in der theorie des spanens von holz. Holztechnologie **2**, 67–70 (1983)
5. Sitkei, G.: Mechanics of Agricultural Materials. Elsevier, Amsterdam (1986)

6. Sitkei, G.: On the mechanics of oblique cutting of wood. In: Proceedings of 13th IWMS Vancouver, pp. 469–476 (1997)
7. Hamilton, G., Goodman, L.: The stress field created by a circular sliding contact. J. Appl. Mech. 371–376 (1966)

Chapter 2
Thermal Loading in Cutting Tools

When wood is processed by machines, the chip slides on the surface of the tool and friction work transforms into heat. Most of this heat flows into the direct surroundings of the edge tool. Depending on the friction power, the area of the edge warms up and the temperature on the surface of the edge can reach a large value.

The high temperature of the edge of a tool increases the wear of a tool as heat weakens and softens metals—especially above 500–600 °C. Microscopic tests of the tools of the timber industry have shown that the smearing of the material is noticeable on the edge surface which is the consequence of the softening of the material. Recently deposits of calcium and potassium were observed on the knife surfaces which can be removed from the wood only at around 800 °C [1, 2].

Although the existence of high temperature on the edge surface is expected [1, 2], no direct measurements or calculations for determining the thermal state of the cutting edge have been presented yet.

In the following we give a detailed analysis of the heat generation in the contact surface, the rate of heat flow into the edge and rake face, the amount of heat divided into the edge and rake face, the amount of heat divided into the cutting edge and the wood and the heat transmission from a rotating tool into the surrounding air. Furthermore, the influence of tool geometry and operating conditions on the thermal loading of the cutting edge is presented. Working diagrams are constructed to facilitate the quick estimation of various influencing factors on the thermal state of the knife.

2.1 Heat Generation in the Contact Surfaces

Tools used in the metal and wood industry differ substantially from each other in one aspect. In cutting metals, the whole chip cross section deforms plastically and the resulting chip thickness is two to three times larger than the thickness of the chip that was set. This plastic deformation is associated with remarkable energy absorption and heat generation inside the chip.

E. Csanády and E. Magoss, *Mechanics of Wood Machining*,
DOI: 10.1007/978-3-642-29955-1_2, © Springer-Verlag Berlin Heidelberg 2013

In cutting wood, we cannot observe a similar occurrence and the cross section of the chip hardly change. Therefore the source of heat is the friction work done on the contact surfaces.

A pressure distribution develops on the edge and front surface of the knife (Fig. 2.1), while the chip moves with v speed in relation to the knife. The theoretical value of the generated heat is:

$$Q = v \int_F \mu\sigma \cdot df \quad \text{(Watt)}$$

The pressure distribution on the knife is not known with sufficient accuracy, therefore we will use the basic relationships of the cutting theory.

The frictional power on the front surface of the knife is determined by the friction force and the sliding speed:

$$Q = \mu N \cdot v$$

where N is the normal force on the rake face.

Combining Eqs. (1.1, 1.7 and 1.8a) we may write:

$$N(\sin\delta + \mu \cdot \cos\delta) = B \cdot h$$

from which

$$N = \frac{B \cdot h}{(\sin\delta + \mu \cdot \cos\delta)} \cong B \cdot h \tag{2.1}$$

as the value of the denominator is close to the unit in the range of $\delta = 60\text{–}70^\circ$ and $\mu = 0.25\text{–}0.3$.

Rotating tools work only in a given rotation angle φ per turns (Fig. 2.2a), heat flows only in this range of angles. The value of the φ angle is calculated as:

$$\varphi = \arccos \cdot \left(1 - \frac{H}{R}\right)$$

Because of the changing cross section of the chip, the equivalent cutting force is smaller than its maximum value (Fig. 2.2b) and their ratio will be given with the parameter:

$$\psi = \frac{P_{he}}{P_{h\max}}$$

Usually $\Psi = 0.8$ value can be taken as a good approximation. The total heat introduced on the rake face is:

$$Q = \mu B h b v \psi \left(\frac{\varphi}{360}\right)$$

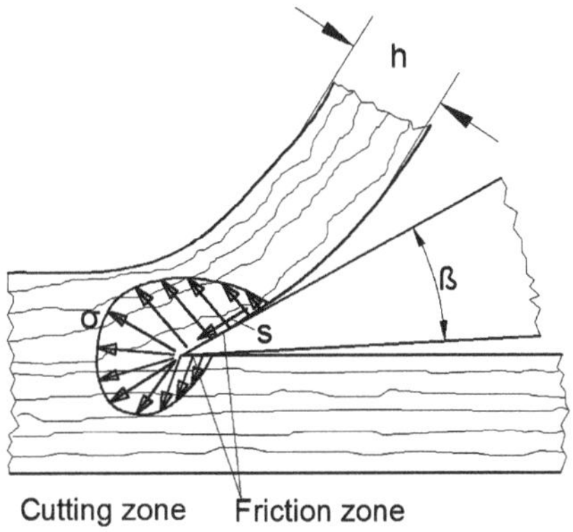

Fig. 2.1 The stress distribution on the surface of the tool [4, 6]

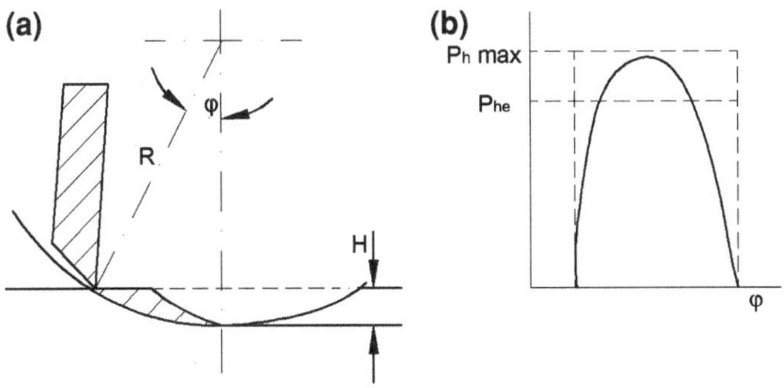

Fig. 2.2 The rotation angle of the tool during cutting (**a**), the course of the cutting force and the determination of the average cutting force (**b**)

The equivalent length of the heat inflow on the rake face depends primarily on the thickness of the chip and its value, from experimental observations and theoretical considerations, may be taken as:

$$L = (1.2 \sim 1.3) \cdot h$$

The theoretical value of the specific heat flow on the rake face can be written as follows:

$$q = \frac{1}{1.3}\mu B b v \psi\left(\frac{\varphi}{360}\right) \quad \left[\frac{W}{mm}\right] \tag{2.2}$$

Taking into consideration the expression of the force acting on the edge (see Eq. 1.8), the theoretical value of the specific heat flux to the edge will be:

$$q_a = \mu \sigma_c b v \psi_1\left(\frac{\varphi}{360}\right) \quad \left[\frac{W}{mm}\right] \tag{2.3}$$

The arc length of the edge:

$$L_a = \rho(\pi - \delta)$$

where

ρ means the radius of the edge

δ is the cutting angle and its value has to be substituted in radian

ψ_1 parameter takes into account a force decrease at the exit point of the knife with $\psi_1 = 0.9$ value

Figure 2.3 demonstrates the change of the specific heat fluxes depending on the *H/R* ratio ($v = 37.7$ m/s). As we can see, the heat flux introduced on the edge is more than double compared to the heat on the rake face.

2.2 Boundary Conditions

The temperature field developing in the knife depends on many factors. The main factors are the following:

(a) the geometric characteristics of the knife, the heat conductivity of the knife material and its specific heat;
(b) the true heat flow is smaller than the theoretical value both on the edge and on the rake face, because friction is an irreversible process. Furthermore, friction, as a shearing stress, generates deformations in both contacting materials, which also consumes energy, and finally some heat will be dissipated into the air,
(c) Intensive heat transfer occurs on the surface of fast rotating tools that partly decreases the amount of heat and cools the knife while rotating freely in the air,
(d) the frictional heat is generated on the contact surface of the chip and the tool, so a smaller part of the heat flows into the chip, the greater part flows into the knife,
(e) a further possible inaccuracy may be given by the fact that the effective friction coefficient between the tool and wood cannot be measured during the cutting process. Friction coefficient values available were determined from cutting force measurements.

The previous questions can be answered partly from experimental results and partly from theoretical considerations. If the boundary conditions (the location of the heat flow and intensity, heat transfer on the surface) are known, then the differential equation of the heat conductivity can be solved with a computer (finite element method) and the whole temperature field can be determined.

The comparison of temperature measurements made inside of the knife and the theoretical calculations showed that only around 60 % of the theoretically

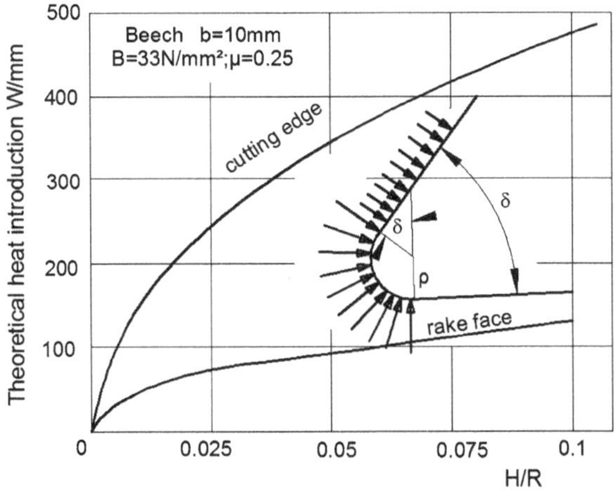

Fig. 2.3 The theoretical distribution of the heat flow on the rake face and on the edge of the tool ($v = 37.7$ m/s) [5, 6]

determined frictional heat appears as effective heat on the frictional surface. In the following calculations, we have taken 60 % of the theoretical values calculated with Eqs. 2.2 and 2.3.

The smaller portion of the generated heat is taken away by the chip. This amount of heat can be estimated as follows. The temperature variation of the infinite half-space applying a constant surface heat flux can be calculated from the following equation [3]:

$$\vartheta = \frac{q_0}{\lambda}\sqrt{\frac{4at}{\pi}}\left[e^{-x^2/4at} - x\sqrt{\frac{\pi}{4at}} \cdot erfc\left(\frac{x}{\sqrt{4at}}\right)\right]$$

where:

$$erfc = 1 - \frac{2}{\sqrt{\pi}}\int_0^z e^{-\beta^2}\,dz$$

λ heat conduction coefficient,
q_0 heat flux on the surface,
t the time,
a temperature conductivity.

The temperature of both materials is the same at the contact surface; therefore, the following equality can be written for the contacting surfaces ($x = 0$):

$$\frac{q_w}{\lambda_w}\sqrt{\frac{4a_w t}{\pi}} = \frac{q_k}{\lambda_k}\sqrt{\frac{4a_k t}{\pi}}$$

from which, the heat ratio is calculated with the following equation:

$$\frac{q_w}{q_k} = \frac{\lambda_w}{\lambda_k}\sqrt{\frac{a_k}{a_w}} \tag{2.4}$$

In the previous equations, the w index is related to the wood, and the k index to the knife. The heat conductivity and the temperature conductivity coefficients of the wood depend on its moisture content. The values of λ and a are summarized in Table 2.1.

Calculations showed that 3.5–10 % of the total heat flows into the chip depending on its moisture content.

This is in agreement with practical observations that cutting wet wood considerably decreases the knife-temperature.

When machining wood with high moisture content, water is pressed out to the surface and the heat needed for evaporation further decreases the surface temperature.

Most of the heat amount flowing into the tool radiates from the surface of the tool into the air. A complicated convective heat transfer develops around the tool, so the heat transfer coefficient can only be reliably determined with experiments. The method is the following: after the tool has reached a steady heat condition, cutting is interrupted and the cooling down curve (see Fig. 2.12) is registered at some points of the rotating tool. Then we calculate the cooling down curve of the point, (finite element method), by using different heat transfer coefficients. When the measured and the calculated cooling down curve coincide with each other, the heat transfer coefficient agrees the real value. The results of these kinds of measurements and calculations are demonstrated in Fig. 2.4. The result can be described with the general similarity equation

$$Nu = 0.12 \cdot Re^{0.7} \tag{2.5}$$

with

$$Nu = \frac{\alpha d}{\lambda} \quad \text{and} \quad Re = \frac{v \cdot d}{\upsilon}$$

where:

Nu Nusselt number,
Re Reynolds number,
v circumferential speed of the tool,
d diameter of the tool,
α heat transfer coefficient,
λ the heat conductivity coefficient of the air,
υ the kinematic viscosity of the air.

It is interesting to note that the similarity equation valid for the plane wall ($Nu = 0.032\ Re^{0.8}$) gives somewhat similar results.

Table 2.1 Thermal parameters

Material	λ, W/m °C		a, m²/h	
	U = 0 %	U = 30 %	U = 0 %	U = 30 %
Wood	0.163	0.36	0.00055	0.00028
Steel	40		0.042	

Note: U is the moisture content, wet basis

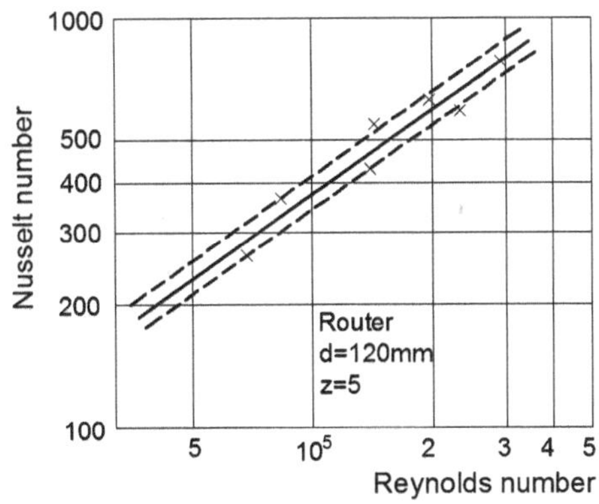

Fig. 2.4 Similarity relationship of the heat transfer criteria for milling tools [4, 6]

2.3 The Temperature of the Tool in Different Operating Conditions

Concerning thermal condition of a tool, the following characteristic operating conditions may be distinguished:

(a) true steady state operating (e.g. wood turning), when the temperature remains constant, independently of time at each point of the tool;
(b) quasi-steady state operating (rotating tools), when the temperature of the tool remains nearly constant, except for the surface and a layer a few tenths of millimetres under it;
(c) unsteady state (not stationary) operating condition at the heat introduction, on and under the surface in a thin layer (in rotating tools see Fig. 2.5);
(d) unsteady state in the whole volume of the tool, the warming up and cooling down period of the tool.

Rotating tools always work in an unsteady heat state. The duration of the inflow of heat is just $\varphi = 15\text{–}25°$ per turn. During the remaining part of the angle of rotation, the surface of the tool edge cools down (Fig. 2.5). Therefore on the surface of the edge the temperature is highly variable around the steady state value.

The amplitude of the temperature variation decreases rapidly as a function of depth. If we approximate the surface temperature variation with the harmonic function

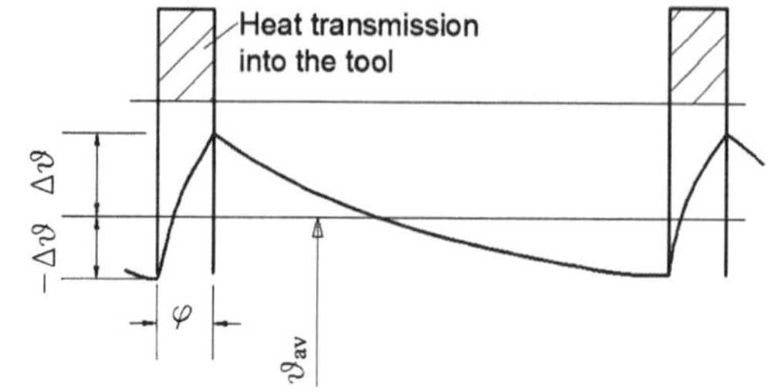

Fig. 2.5 Periodic heat inflow and temperature variation in rotating tools [6]

$$\vartheta = \vartheta_0 \cos\left(\frac{2 \cdot \pi \cdot n \cdot t}{t_0}\right)$$

then the temperature variation under the surface, depending on the depth and the time can be calculated from the following expression [3]:

$$\vartheta_{x,t} = \vartheta_0 e^{-\sqrt{K} \cdot x} \cos\left(\sqrt{K}x - \frac{2 \cdot \pi \cdot n \cdot t}{t_0}\right),$$

where:

$$K = n\pi / at_0$$

and

n 1, 2, 3,…
t_0 the time of one revolution,
a the temperature conductivity.

The amplitude will be the greatest, if the cos-term value is one. In this case the decrease of the temperature amplitude as a function of depth (Fig. 2.6) will be calculated as

$$\frac{\vartheta_{x,t}}{\vartheta_0} = e^{-\sqrt{K} \cdot x}$$

or

$$x = \sqrt{K} \cdot \ln\left(\frac{\vartheta_0}{\vartheta_{x,t}}\right) \qquad (2.6)$$

Taking a rotation speed of 6,000 rpm, for which $t_0 = 0.01$ s, to a 1/10 or 1/100 amplitude ratio will correspond to $x = 0.44$ or 0.88 mm depth, respectively. Due to the short heat inflow, the real x values are smaller, as the calculations with the finite element method have shown. We concluded that the cyclic fluctuation of the temperature around the site of heat inflow can be neglected at a depth of 0.5 mm.

The steady temperature of rotating tools primarily depends on the thickness of the chip, the edge rounding radius, the sharpening angle β and the peripheral

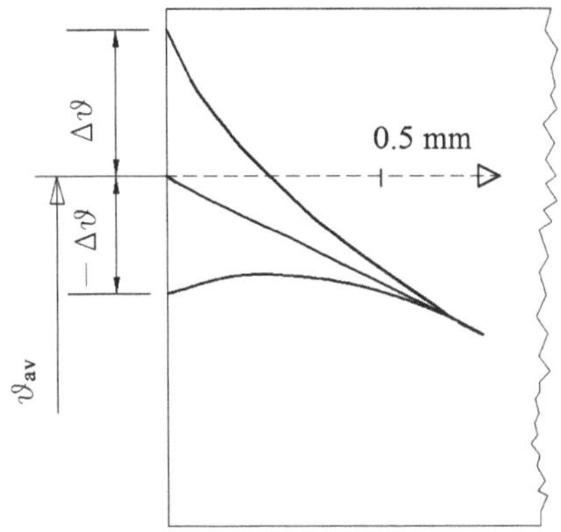

Fig. 2.6 Decrease of the temperature amplitude depending on the depth [3]

speed. Less influencing factors are the strength properties of the wood and the moisture content, the arc length of the cutting, the overload of the gullet with sawdust etc.

The calculations with the finite element method showed that the maximum temperature always occurs on the surface of the edge, approximately at the intersection point of the bisector and the circumference of the edge. Next we review the basic relationships of this maximum temperature as a function of the influencing factors. All measurements and calculations refer to HSS steel.

Figure 2.7 shows the variation of the maximum temperature depending on the rotation speed ($d = 120$ mm). The parameter x notates the heat inflow length on the rake face, which is proportional to the thickness of the chip ($x = 1.2 \sim 1.3\ h$). The maximum temperature increases by the 0.8 exponent of the rotation speed or peripheral speed.

Figure 2.8 shows the maximum temperature depending on the heat inflow length for different edge rounding radii (different sharpness). As we can see, the edge wear substantially increases the maximum temperature.

Figure 2.9 shows measured and calculated maximum temperatures as a function of feed per tooth [4, 5]. Measurements were taken in the bisector of the edge and a thin thermocouple was located at $\Delta x = 3.7$ mm from the edge surface. The empirical equation describing the temperature variation has the form:

$$\vartheta_x = \vartheta_0 + [180 + 465 e_z^{0.5}] e^{-0.3\Delta x}$$

where ϑ_0 means the ambient temperature. Obviously on the surface $\Delta x = 0$ holds.

The sharpening angle β decisively influences the conduction of heat to the air; therefore, the temperature depends strongly on the β angle. In Fig. 2.10 we can see the variation of the maximum temperature for different sharpening angles as a function of the length of the heat inflow.

The curves of the Figs. 2.7, 2.8 and 2.10 can be described with the following general equation:

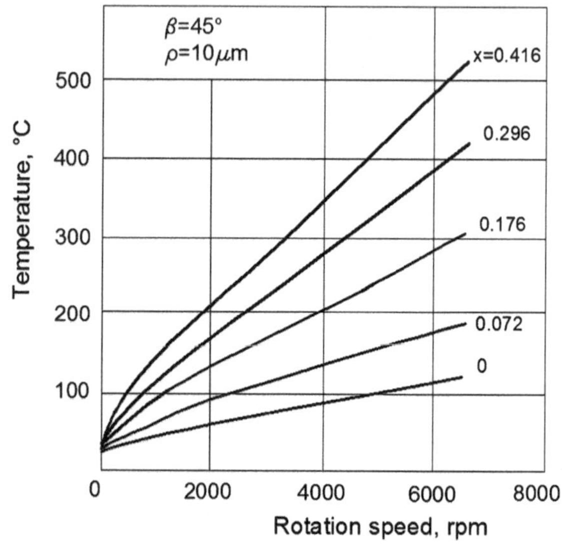

Fig. 2.7 The maximum temperature of the edge surface depending on the rotation speed for different length of heat inflow on the rake face (x in mm) [5, 6]

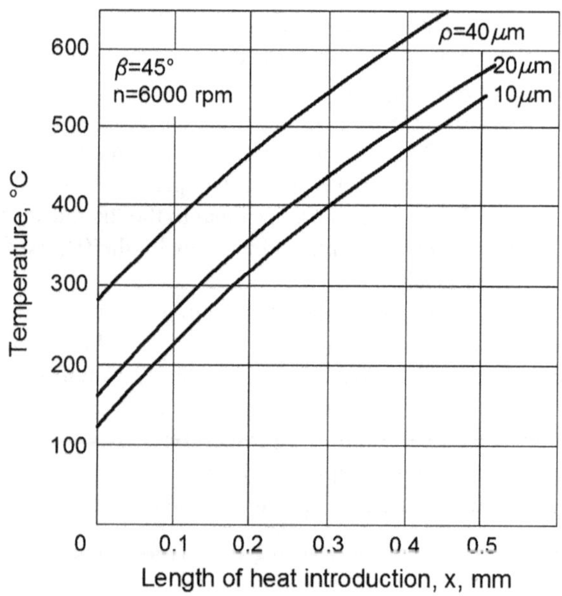

Fig. 2.8 The effects of edge wear on the maximum temperature for different length of heat inflow [5, 6]

$$\vartheta_{\max} = \vartheta_0 + \left[26.6\rho^{0.8} + 1503 \cdot x^{0.8}\right] \frac{v^{0.8}}{\beta^{0.95}} \tag{2.7}$$

where: ϑ_0—is the ambient temperature and the value of ρ is in μm, x is in mm, v is in m/s and β should be substituted in degrees. In Fig. 2.11 we can see experimental results of circular saws ($d = 400$–500 mm) measured at the bisector of the teeth at

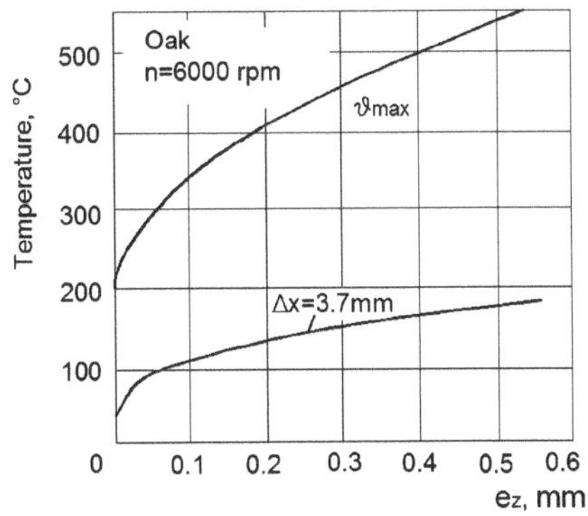

Fig. 2.9 Measured and calculated maximum surface temperatures and temperatures under the edge as a function of feed per tooth [4, 5]

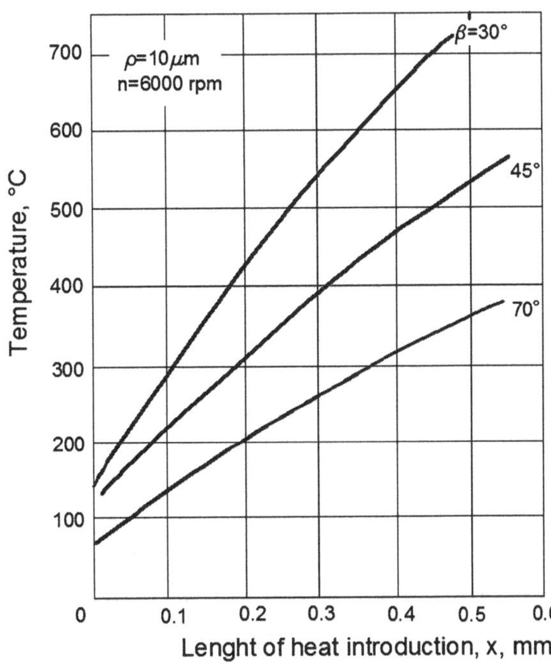

Fig. 2.10 The effect of a sharpening angle on the maximum edge temperature [4, 5]

a distance Δx from the edge [6]. The extrapolation of the measured points until $\Delta x = 0$ gives the edge surface temperature and these temperature values can be described well with Eq. 2.7. The temperature of any points on the bisector can be given with the following equation:

$$\vartheta_x = \vartheta_{\max} \cdot e^{-k \cdot \Delta x^n}$$

Using the data from Fig. 2.11, we get the following relationship ($\beta = 55^\circ$ and $\rho = 20$–22 μm):

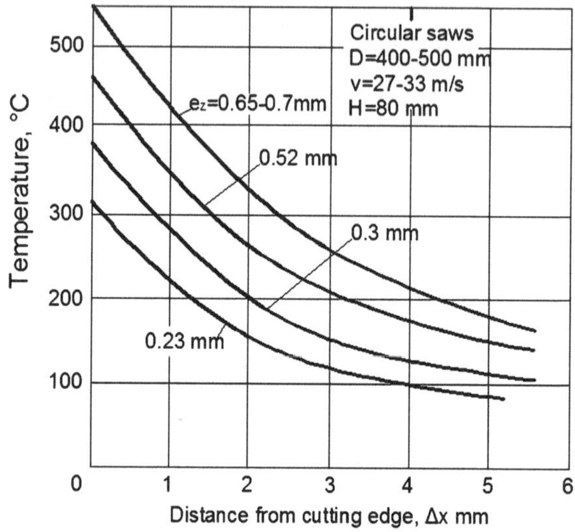

Fig. 2.11 Variation of the temperature in the bisector of the edge for circular saws working with different feeds per tooth [6]

$$\vartheta_x = \vartheta_0 + \left[7 + 586\left(\frac{\varphi}{360}\right)e_{zav}^{0.8}\right] \cdot v^{0.8} \cdot e^{-0.4\Delta x^{0.65}} \tag{2.8}$$

where e_{zav} means the average feed per tooth in mm.

The typical operating condition of cutting tools is the warming up and the cooling down periods. In Fig. 2.12 we can see typical warming up and cooling down curves for a milling cutter at five different points located on the bisector [4, 5]. Both the warming up and the cooling down are relatively fast, the time constant chiefly depends on the distance from the edge. (Time constant is defined as the time required to reach 63.2 % of the maximum temperature increase.)

Using circular saws, the length of the cutting arc, and also the energy of friction forces influences the heat inflow. The length of the cutting arc related to the cutting height *($\varphi R/H$)* is a measure for the contribution of friction forces (see Fig. 3.11).

The warming up and cooling down of the edge surface can be determined only with calculation. These kinds of calculated results are shown in Fig. 2.13 as a function of time. The time constant of the edge surface is approximately 2 s. Moving away from the surface, the warming up and the cooling down take more time, and the time constant increases considerably. The temperature of the tool itself changes slowly and only slightly at 3–4 cm away from the edge surface.

Rotating tools cut only for a short time within one revolution. Therefore a cyclical temperature variation occurs on the edge surface that decreases quickly as a function of depth. Those calculations are shown in Fig. 2.14, showing the temperature variation of the first three cycles depending on the angle rotation. The amplitudes are very large, and after reaching the steady state temperature (400–500 °C) the peak temperature can spike to 1,000 °C for an instant.

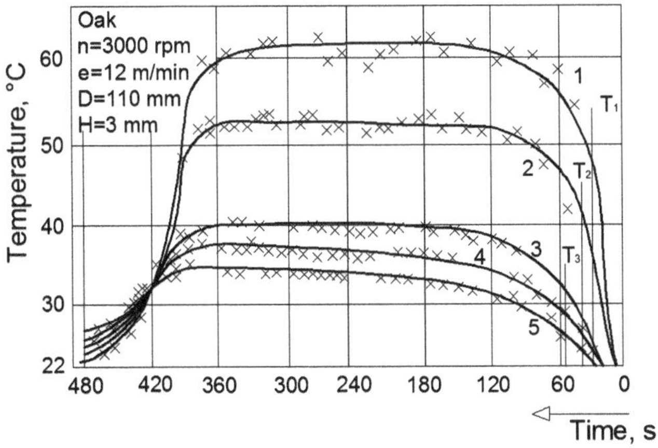

Fig. 2.12 The warming up and cooling down of a tool measured in the bisector and in different distances from the edge surface [4, 5]

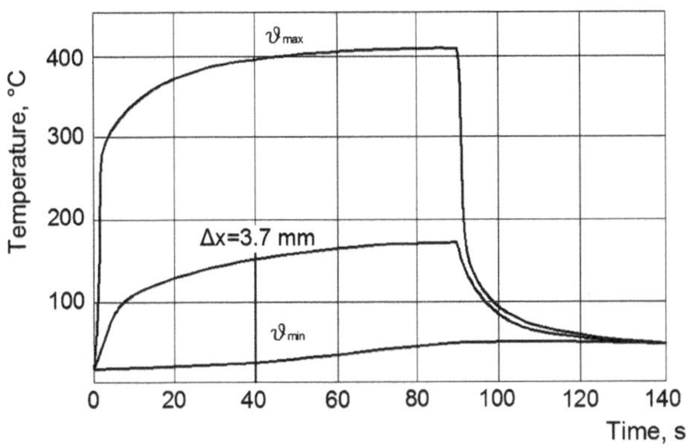

Fig. 2.13 Calculated warming up curves at different points of a tool [4, 7]

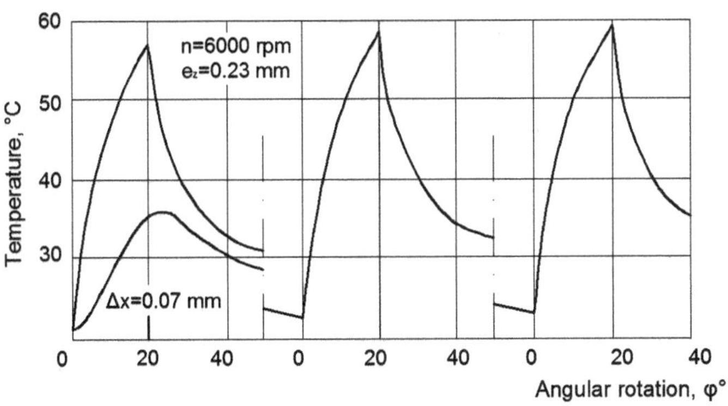

Fig. 2.14 The first three cycle of an unsteady warming up process as a function of angular rotation [5, 7]

From the previous discussion the following conclusions can be drawn:

- the heat load in woodworking tools primarily originates from frictional heat;
- the high peripheral speed generates large frictional power and, as a consequence, high surface temperatures;
- the high surface temperature is an important cause of wear, as the materials soften and high temperature corrosion occurs (see in Chap. 7);
- the high cyclic temperature variations of the edge surface causes cyclical heat-stress that can cause pitting of the surface;
- the edge temperature of a worn tool is always larger than for a sharp tool which further increases the intensity of the wear.

References

1. Демяновский, К., Износостойкость инструмента для фрезерования древесины, (Wear resistance of tools for woodworking), Изд. Лесная Пром. (1968)
2. Stewart, H.: Feasible high-temperature phenomena in tool wear. For. Prod. J. pp. 25–28, March (1989)
3. Carslaw, H., Jaeger, J.: Conduction of heat in solids. Oxford University Press, New York (1959)
4. Csanády, E., Faforgácsoló szerszámok hőterhelése: Thermal load in wood cutting tools. PhD Thesis, University of Forestry and Wood Sciences, Sopron, Hungary (1993)
5. Sitkei, G., Csanády, E.: Thermal loading in wood cutting tools. Proceedings of the 1st conference on the development on wood science, pp. 359–368, Buckinghamshire, UK (1994)
6. Sitkei, G. et al.: Theorie des Spanens von Holz. Fortschrittbericht No.1. Acta Fac. Ligniensis Sopron (1990)
7. Csanády, E.: Thermal load in wood cutting tools. Proceedings of 17th IWMS, pp. 28–43, Rosenheim (2005)

Chapter 3
Operating Parameters of Wood Cutting Tools

It is important to operate woodworking machines and tools economically and correctly. The optimal operating parameters and manufacturing costs change depending on the task. The operational parameters also have limits. Exceeding these limits may hamper the safe operation of the machine or the tool.

The most important operating parameters are the following:

- the peripheral speed of the tool,
- the feed speed,
- the feed per tooth,
- the depth of cut.

The operational parameters influence the following for a given tool:

- the kerf volume cut in the unit time,
- the cross section cut in the unit time,
- the surface roughness and waviness,
- the wear of the tool,
- the energy consumption of cutting,
- the costs of machining.

The ultimate goal of any woodworking operation should be to minimize energy consumption and manufacturing costs, and to maximize throughput and quality.

In the next section, we give the most important relationships of cutting tools used in woodworking operations.

3.1 The Bandsaw

The bandsaw has a straight and uniform motion during work, while the material is pushed against the saw perpendicular to the blade. The kinematic relations and notations are shown in Fig. 3.1. The parameters of the saw blade are its thickness (s),

E. Csanády and E. Magoss, *Mechanics of Wood Machining*,
DOI: 10.1007/978-3-642-29955-1_3,

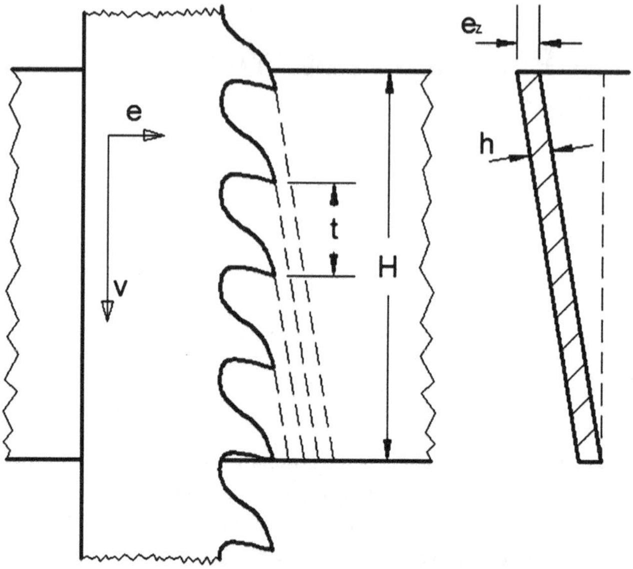

Fig. 3.1 The kinematic relations of the bandsaw [1]

the extent of its spring set or the swage set (*2a*), the volume of its gullet (V_g) and the pitch of its teeth (*t*).

The speed of the cutting is practically the same as the band speed, as the feed speed is relatively small. The time needed to pass the length of the tooth-pitch is

$$\tau = \frac{t}{v}$$

during this time the material is pushed forth with e_z value,

$$\tau = \frac{e_z}{e}$$

and the feed per tooth will be

$$e_z = \frac{e}{v} t. \tag{3.1}$$

The v speed of the tool is the same as the peripheral speed of the band mill wheel

$$v = \frac{D \cdot \pi \cdot n}{60}$$

where:

D is the wheel diameter and *n* is the rotation speed.

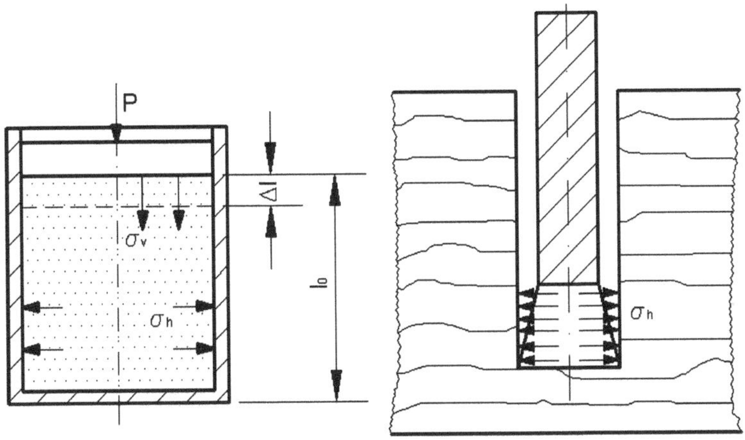

Fig. 3.2 The compaction of the chip and the formation of lateral pressure in the gullet [1]

The solid wood volume cut by one tooth with $b = s + 2a$ width is calculated by

$$V = l \cdot h \cdot b = e_z H \cdot b = \frac{e}{v} H \cdot t \cdot b$$

where:
l the length of chip,
H the height of the cut cross section.

The chip and solid wood volume ratio is generally $\varepsilon = 3–4$ related to the solid wood, and with this the loose chip volume due to woodworking is:

$$V_{chip} = \varepsilon \frac{e}{v} H \cdot t \cdot b \quad (3.2)$$

The chip has to fit in the gullet until its exit from the material, therefore $V_g = V_{chip}$.. Therefore the value of the maximum theoretical feed speed is from Eq. (3.2):

$$e_{theo} = \frac{V_g \cdot v}{\varepsilon \cdot H \cdot t \cdot b} \quad (3.3)$$

where V_g means the gullet volume.

We can define the *gullet feed index* as the ratio of the loose chip volume and the gullet volume (gullet area x kerf width).

The measurement of compaction of the chip has shown that only 0.2–0.4 bar pressure is needed for a 20–25 % compaction. As a consequence, in practice the theoretical value of the feed speed can be increased by 25 %. Figure 3.2 shows the mechanism of compaction and the development of the lateral pressure in the gullet. The lateral pressure coefficient changes between 0.25–0.35 in static pile of chips. Due to the movement of chips in the gullet, greater values are expected in the gullet than in the static case. If a greater amount of chips are forced into the gullet

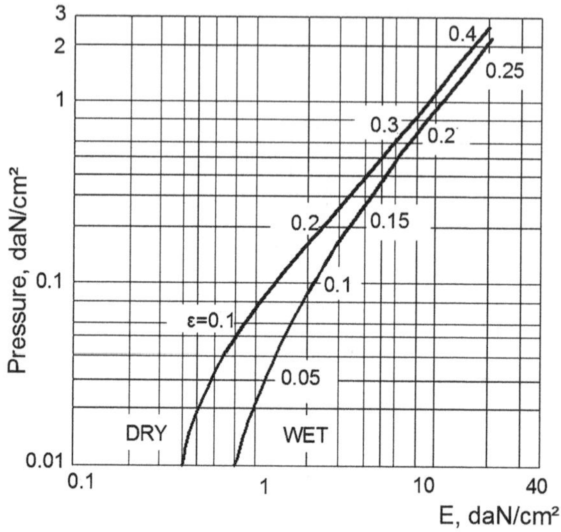

Fig. 3.3 Compaction of saw chips in dry and wet conditions, showing the corresponding strain values [1]

than the allowable, then the increased lateral pressure generates large frictional forces, raising the temperature of the tooth while the cutting force increases.

Figure 3.3 shows compaction measurements on saw chips in dry and wet conditions [1]. The corresponding strain values are also given.

The band saw can be spring set or swage set. A swage set tooth makes a better surface and its cutting width is smaller compared to the spring set. The stability of a swage set tool is also better. Therefore the sawing deviation of long straight cuts is much smaller.

3.2 The Frame Saw

The frame saw has a reciprocating movement and, because of the finite length of the connecting rod, the movement differs from the harmonic motion. The vertical displacement of the frame can be calculated from the following equation depending on the rotation angle (Fig. 3.4):

$$x = r\left(1 - \cos\varphi + \frac{\lambda}{2}\sin 2\varphi\right) \tag{3.4}$$

where:

r crank radius

φ rotation angle,

l the length of the connecting rod,

λ $= r/l$ the ratio of the crank radius and the length of the connecting rod.

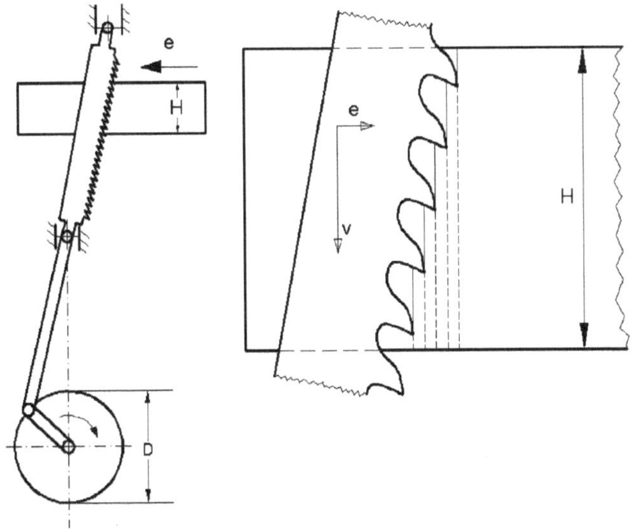

Fig. 3.4 The kinematic relations of the frame saw [1]

The continuous feed of the material can be possible in that way that the frame is inclined forward with κ angle, so by the moving upwards of the frame it makes a way free in horizontal direction:

$$y_1 = x \cdot tg\kappa$$

The feed has a constant speed, and the corresponding displacement is

$$y_2 = e \cdot \tau$$

It follows that there is a close relation between the inclination of the frame and the feed speed. The following equations can be written for a half turn ($\varphi = 180°$):

$$\tau = \frac{\varphi}{6n} = \frac{30}{n} \text{ and } \tau = \frac{y_1}{e} = \frac{2r \cdot tg\kappa}{e},$$

and equating the above two equations yields

$$e = \frac{2r \cdot n \cdot tg\kappa}{30} \quad (3.5)$$

The feed in one turn ($\varphi = 360°$)

$$e_n = e \cdot t = 4r \cdot tg\kappa \quad (3.6)$$

The average value of the feed per tooth

$$e_z = e_n \frac{t}{2r} = 2t \cdot tg\kappa \quad (3.7)$$

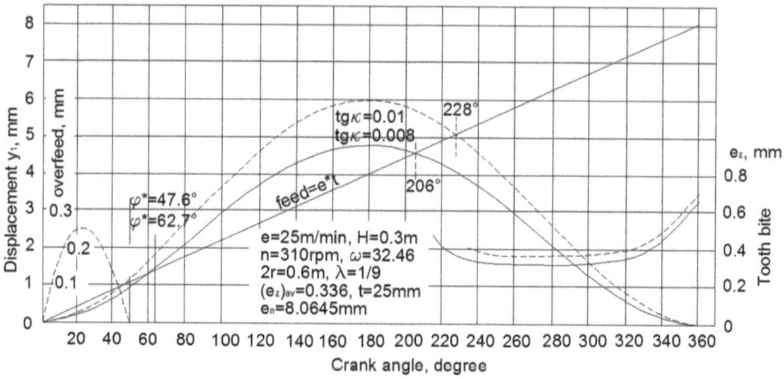

Fig. 3.5 Working diagram of a frame saw for two different forward inclinations, variation of tooth bite and overfeed as a function of the angle of rotation [1]

Figure 3.5 (indicator diagram) represents the kinematic relations of the frame saw. If we change the inclination under the same conditions, then the sawing begins after the upper dead point at different angle. It changes the effective feed per tooth and the extent of the overfeed after the lower dead point. The over feed phenomenon always occurs because the movement of the frame is slow around the dead point, while the feed speed is constant.

One of the characteristic features of the frame saw is the dynamics of logs passing the bottom dead centre. This problem is of kinematic nature, i. e. the frame stops at the bottom dead point, while the feed of the log remains continuous. As a consequence, the log will be pushed onto the saw blades producing dynamic load in the blade and locally on the edges. This dynamic load depends on the feed rate, the pitch of teeth and on the forward inclination of the frame.

In order to reduce these dynamic loads, some manufacturers offer their machines with a feed interrupting mechanism built into the hydraulic drive of the feed rolls. However, the interrupting mechanism must be accurately adjusted; otherwise there is a considerable feed loss.

The tooth tip trajectories are needed in order to determine the overfeed values. The path of a tooth tip within the log can be calculated and plotted as shown in Fig. 3.6 [2].

The overfeed values depend on the pitch of the teeth and the feed per revolution. This relationship is given in Fig. 3.7. The overfeed values are relatively large and, therefore, the back face of the tooth cannot deform the wood in a measure of the overfeed value. As a consequence, the log must be stopped at the bottom dead centre. Stopping the log will cause an increased slip between the feed rollers and the log.

The above statement was experimentally verified. Figure 3.8 shows the log motion as a function of crank angle. Just after the bottom dead centre, the log stops and this interval is 50–60° of the angular rotation. That means that the pushing force is not enough to deform the wood to an extent which would be equal to the overfeed value.

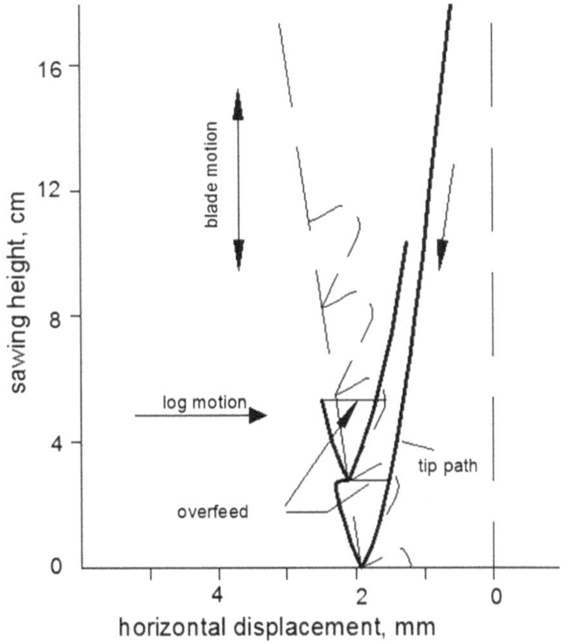

Fig. 3.6 Tooth tip trajectories to determine overfeed values [2]

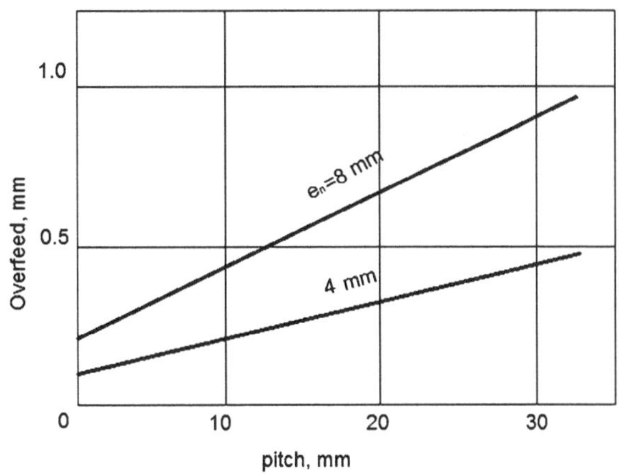

Fig. 3.7 Overfeed values as a function of pitch and feed per revolution [2]

The measurement of maximum pulling forces at the feed rolls gave values between 7,000 and 8,500 N. The greater values are valid for dry logs without bark. With 6–8 blades, this force exerts a load per tooth of around 150 N. Some additional loads originate from the deceleration of the log. This load per tooth is 40–60 N.

The slip between feed rolls and the log depends on many factors such as the wear of ribs and thorns of the feed rolls, the feed speed, the dullness of the saw teeth, the log surface (bark) and its moisture content.

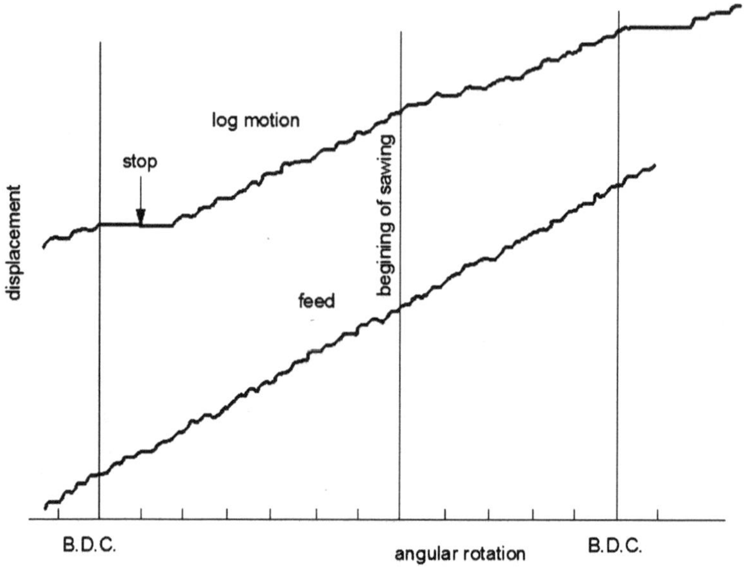

Fig. 3.8 Log motion as a function of crank angle with continuous feed [2]

During upward motion of the blades, no counter-forces act on the log (except a short distance near the bottom dead centre) and in this case the slip is the smallest possible and varies around 10 % (Fig. 3.9). During cutting the slip increases due to cutting forces acting as counter-forces. The total slip related to a complete revolution includes the stopping of the log at the bottom dead centre.

The chip has to fit in the gullet when using frame saw. The chip volume can be calculated similarly, as with a band saw but we have to take into consideration that the frame saw cuts only per half-turns. As a consequence, the chip volume amounts to:

$$V_{chip} = 2 \cdot \varepsilon \frac{e}{v_k} H \cdot t \cdot b,$$

where the average blade speed is

$$v_k = \frac{D \cdot n}{30}$$

With this the theoretical feed speed is:

$$e_{th} = \frac{V_g \cdot v_k}{2\varepsilon \cdot H \cdot t \cdot b} \tag{3.8}$$

Due to the compressibility of chips, the theoretical feed speed can be increased by 25–30 % using swage set teeth. Using spring set teeth a part of the chip can get out of the gullet, and so it relieves the gullet. Then the theoretical feed speed can be nearly doubled, depending on the extent of the spring set.

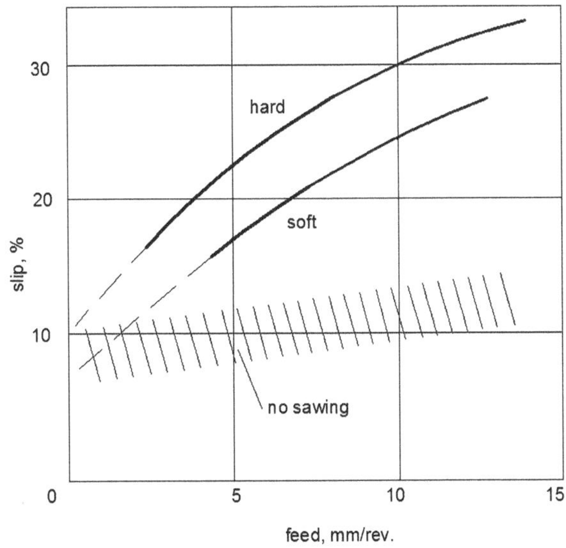

Fig. 3.9 Slip variation between feed rolls and logs for soft and hard woods [2]

3.3 The Circular Saw

Figure 3.10 demonstrates the kinematic relations of the circular saw. The tooth of the circular saw describes a cycloid track in relation to the material. The displacement is described by parametric equations in following manner:

$$x = R \cdot \sin \omega\tau + e \cdot \tau$$

$$y = R \cdot \cos \omega\tau$$

where:

τ the time,
R the radius of the circular saw,
ω the angular speed.

The value of the feed per tooth is

$$e_z = \frac{e}{n \cdot z}$$

where:

z is the number of teeth.

The maximum value of chip thickness is approximately

$$h_{\max} = e_z \cdot \cos \alpha$$

The average chip thickness depends on the H height and the position of the material. For the same H cutting height the φ_O cutting rotation angle will be larger if the material is placed higher over the centre of rotation (distance b in Fig. 3.10).

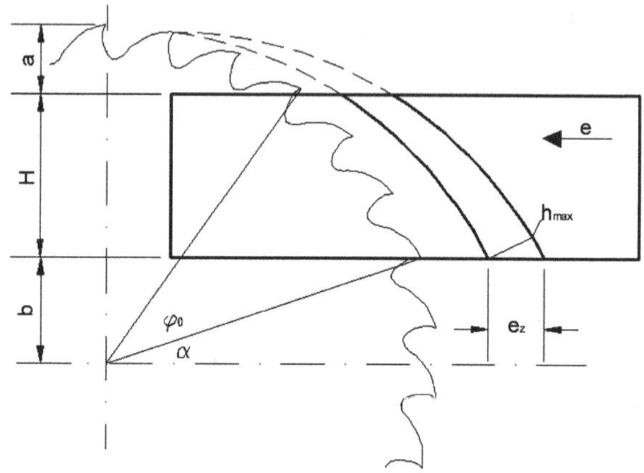

Fig. 3.10 The kinematic relations of the circular saw [1]

The average chip thickness is

$$e_{zav} = \frac{e_z \cdot H}{\varphi_0 R} \tag{3.9}$$

where the angle of the cutting φ_o must be substituted in radian. The angle φ_o can be calculated as follows

$$\varphi_o = \arcsin\left(\frac{b}{R} + \frac{H}{R}\right) - \arcsin\left(\frac{b}{R}\right)$$

It can be stated that the angle φ_o depends on the position of the workpiece in relation to the top of the saw blade. The relative cutting arc influences the friction work of the sawing process. Therefore, the use of a lower machine table is preferable from energetic point of view.

The relative cutting arc is given by the following equation:

$$L_c = \frac{\varphi_0 R}{H}$$

which is illustrated in Fig. 3.11 for different machine table positions as a function of relative cutting height.

It is important to minimize the breaking of the edges of the workpiece (splintering). The direction of cutting forces has a great influence on edge quality. Compression forces provide a better cutting process without edge breaking; therefore, tension forces near the edge should be avoided. Edge breaking can be minimized for the upper or lower edge or both. According to experiments determined that the optimum overhang of the blade rim over the workpiece is around 10 mm as shown in Fig. 3.12, [3].

Circular saws are usually very noisy. Its noise can be reduced with the decreasing of the speed and the feed per tooth, and by increasing of the number of working teeth. It is also effective to coat the surface of the tool with cushioning material, but this raises the cost of the tool.

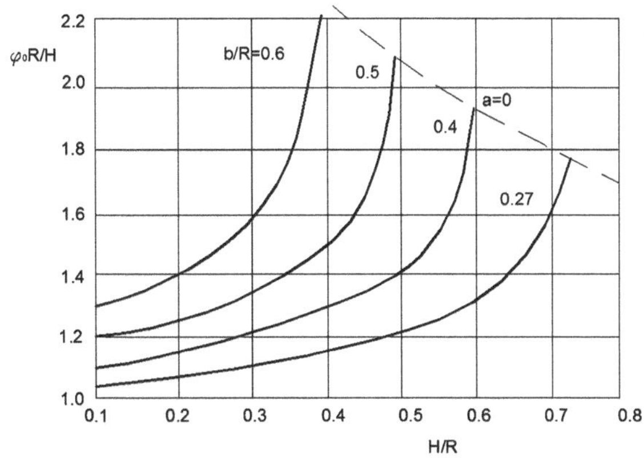

Fig. 3.11 Length of the relative cutting arc for different machine table positions

3.4 Milling and Planing Machines

The kinematic relations of the milling and planing machines are demonstrated in Fig. 3.13. The tool edge describes a cycloid path in relation to the material similar to circular saws.

The value of the feed per tooth is:

$$e_z = \frac{e}{n \cdot z}$$

The maximum chip thickness is approximately:

$$h_{\max} = e_z \cdot \sin \varphi$$

The cutting rotation angle φ—as shown in Fig. 3.14—depends on the H/R ratio. The expression for the φ function is given by:

$$\varphi = 1{,}425\sqrt{\frac{H}{R}}$$

so the average chip thickness:

$$e_{zav} = \frac{e_z \cdot H}{\varphi \cdot R} = \frac{e_z}{1{,}425}\sqrt{\frac{H}{R}} \tag{3.10}$$

Cutting can be climb cutting or counter cutting in relation to the direction of the feed (Fig. 3.15). With climb cutting the horizontal component of the cutting force is in the same direction as the feed. Therefore, the material can be forced under the tool if the restraining force of the feed is not sufficient. For this reason, it cannot be used for a manual feed. With counter cutting, the horizontal component of the cutting force is opposite to the direction of the feed.

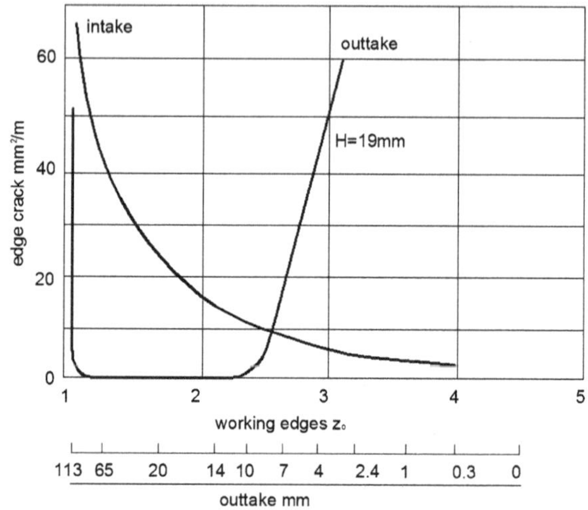

Fig. 3.12 The effect of the tool overhang on the edge breaking for particle board 19 mm thick [3]

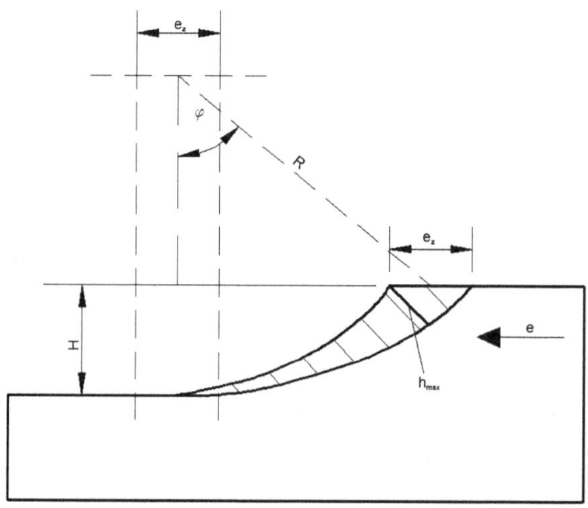

Fig. 3.13 The kinematic relations of a milling machine [1]

Climb cutting generally produces a smoother surface, because the edge of the knife always exerts compressive stress on the material. With counter cutting, the edge exerts tensile stress on the material that can cause grain laceration fibre. But the great cutting speed of counter cutting due to the inertia forces generally produces the necessary surface quality.

The inherent disadvantage of the milling and planing operation is the wavy surface caused by the cycloid path (Fig. 3.16). Factors that determine waviness are the wave depth (t_1) which is determined by the diameter of the blade, number of teeth and rotation speed, and the feed speed of the material.

According Pythagoras' theorem, the wave depth can be calculated as follows:

$$(e_z/2)^2 = (D/2)^2 - (D/2 - t_1)^2$$

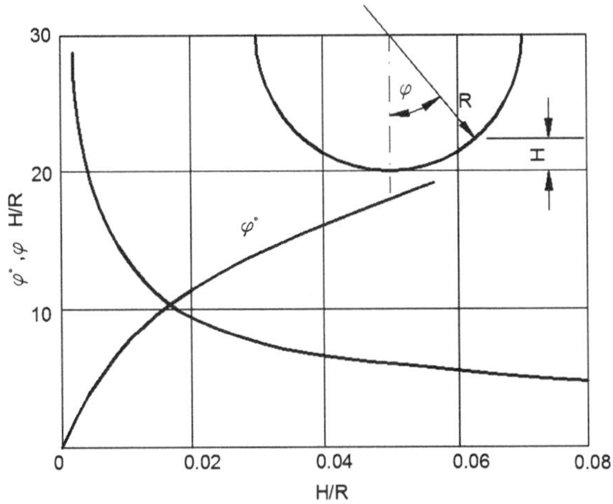

Fig. 3.14 The relative length of the cutting arc depends on the cutting depth [1]

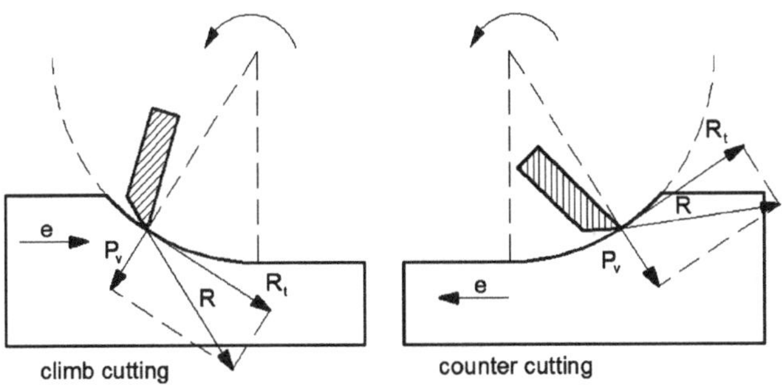

Fig. 3.15 Climb and counter cutting

from which

$$e_z = 2\sqrt{t_1(D - t_1)} \cong 2\sqrt{Dt_1};$$

and

$$t_1 = \frac{e_z^2}{4D} = \frac{e^2}{4Dn^2z^2};$$

From the above equation, the following $D/_{t_1}$ dimensionless number can be constituted:

$$\frac{D}{t_1} = \frac{4D^2n^2z^2}{e^2},$$

taking the square root of the right side

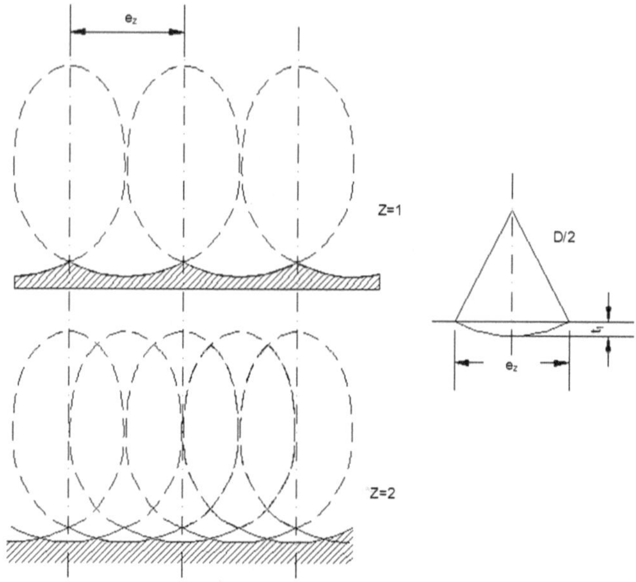

Fig. 3.16 The cycloid path of the milling and planing knife, the development of the waviness [1]

$$\Pi = \frac{Dnz}{e}$$

we get this dimensionless number. The larger this number is, the smaller the waviness. In order to obtain a high quality surface, one has to work in the range of $\Pi = 120-180$.

The 1–2 μm wave depth is high quality, while the 3–6 μm wave depth means a medium quality surface.

In the last decades, several face milling methods were analysed and a new face milling and moulding technique was designed and developed [4–6].

The different moulding systems for edge working are classified and shown in Fig. 3.17. The classical method is the use of the moulding cutter head with peripheral blades. There is a 90° angle between the cutter head axis and the plane of the workpiece. The sawing method (hogging saw) is also often used where the tool axis is parallel with the workpiece plane (plane moulding). These two systems may be considered as extreme concerning the setting angle κ. Between these two extreme values, different angles can be used as shown in Fig. 3.17, [4, 5].

The cutting course varies in different ways depending on the combination of angles which may be ellipsoidal, parabolic or hyperbolic. This last method using a hyperbolic path is very promising to obtain good surface and edge quality and a long service life [5, 6].

The interaction of the workpiece and knife is shown in Fig. 3.18. The path of the principal cutting edge (face cutting edge) is a hyperbolic curve producing a very thin chip at the exit from the wood. The secondary cutting edge (peripheral

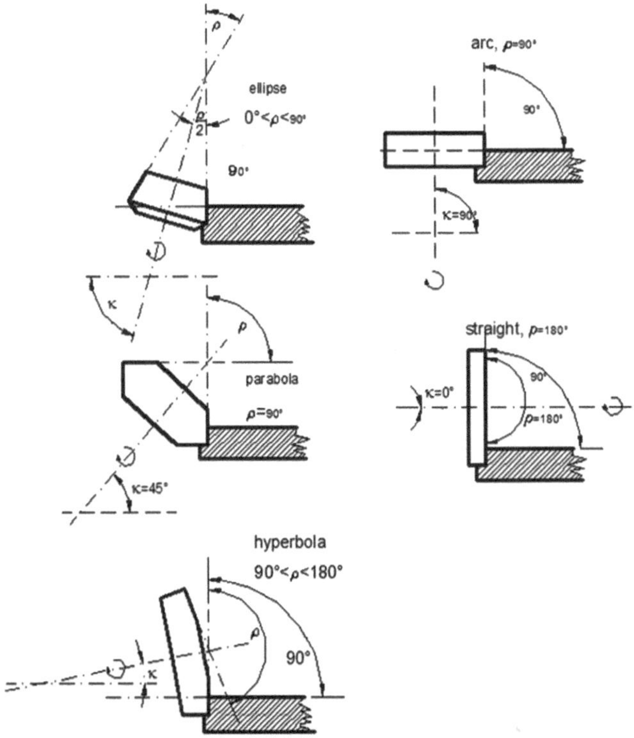

Fig. 3.17 Different moulding systems [4, 5]

cutting edge) will cut a part of the maximum depth of the cut. There may be more splintering of the edge at the entry of the knife into the wood, but this break-out will be eliminated by the next cutting path.

The hyperbolic path of the cutting edge may have an enormous cutting circle radius larger than the tool depending on the tool setting angle κ as shown in Fig. 3.19.

The cutter head has a setting angle κ and a tilting angle λ in the forward direction, as shown in Fig. 3.20. The tilting angle λ results in the above mentioned hyperbolic path of the knife and formation of thin chips at the exit from the wood. This is essential for a good surface and edge quality. It is also important that the cutting circle will be much larger than the diameter of the cutter head [6]. The larger cutting circle provides more compression forces, which may contribute to less break-out from the edge of the workpiece. The optimum choice of the setting and tilting angles can only be established by experiments. The experimental results are shown in Fig. 3.21 where the edge splintering has minimum values between 5 and 10 degrees of the setting angle. In these experiments the cutter head operated with 1 knife and had a diameter of 200 mm. Furthermore, it is remarkable that the knife can work with a tooth feed of 5 mm with almost no edge break-out.

Fig. 3.18 The interaction of the workpiece and the knife [4, 5]

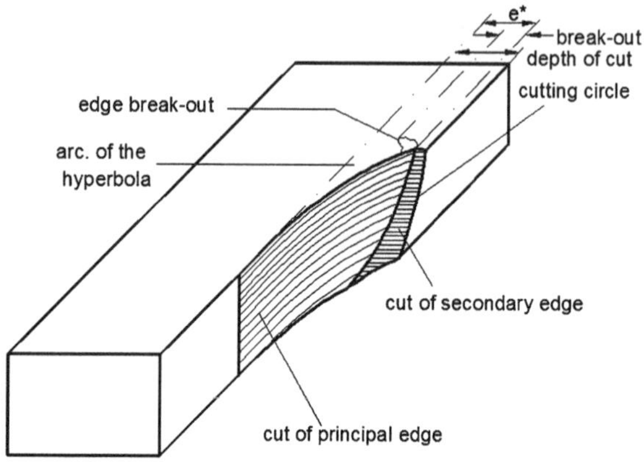

Fig. 3.19 Radius of the hyperbola arc on the workpiece as a function of the tool setting angle (virtual cutting circle)

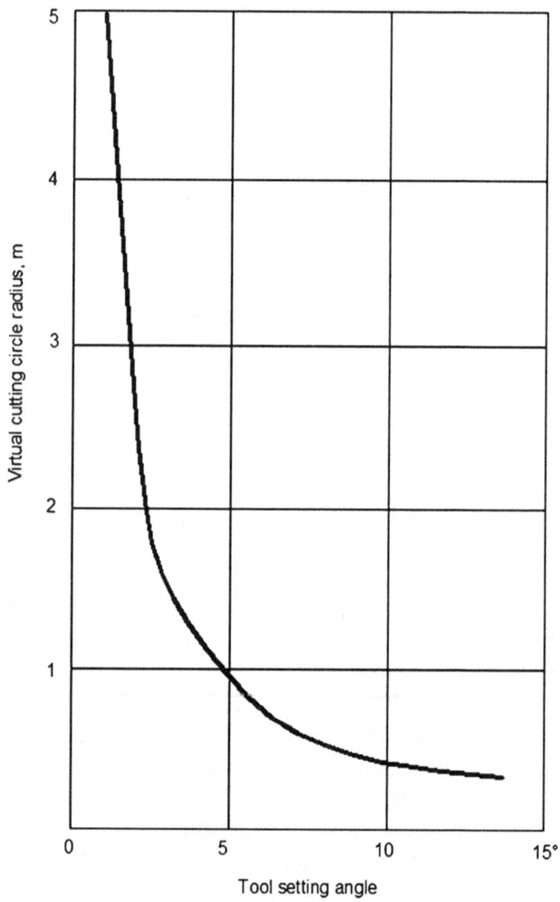

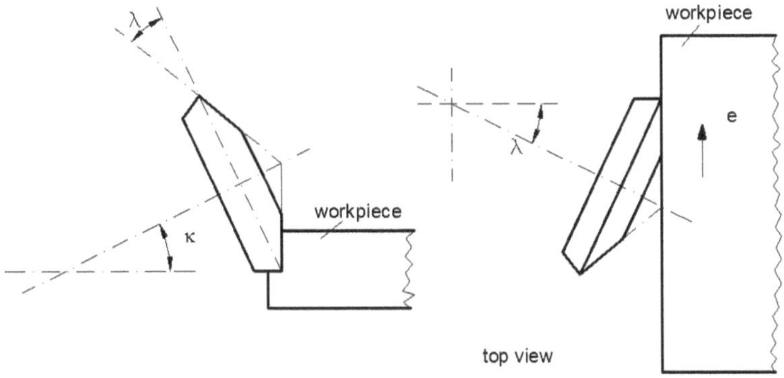

Fig. 3.20 The setting and tilting angles [4, 5]

Having the optimum setting angle range, the next question is to examine the influence of the tilting angle on the quality. The influence of tilting (machine) angle on the edge splintering is shown in Fig. 3.22. Quite a small angle may be selected and the optimum values are from 0.5 to 1.5°. A negative value of λ rapidly increases the splintering of the edge.

Further experiments have shown that the rake angle of the knife may be as low as 5–10°; therefore, TC reversible knives may also be used which makes the system generally applicable. This reduces the wear of the knives. A cutter head with 12 knives equipped with fine grained TC reversible plates has machined 36,000 m long edge surface without needing re-sharpening.

3.5 Drills

Drills are used to make through-holes and dead-end holes mainly in the furniture industry. The main types of drills are seen in Fig. 3.23a–d. Double-edge twist drills are used to drill through holes. They have a diameter d and a cone angle 2κ. The feed per revolution is given by

$$e_n = \frac{e}{n}$$

where e is the feed rate, m/min, and n is the rotation speed, rpm. The feed per tooth will be calculated as

$$e_z = \frac{e}{z \cdot n}$$

where z is the number of cutting edges.

The length of the cutting edge is

$$b = \frac{d}{2 \sin \kappa}$$

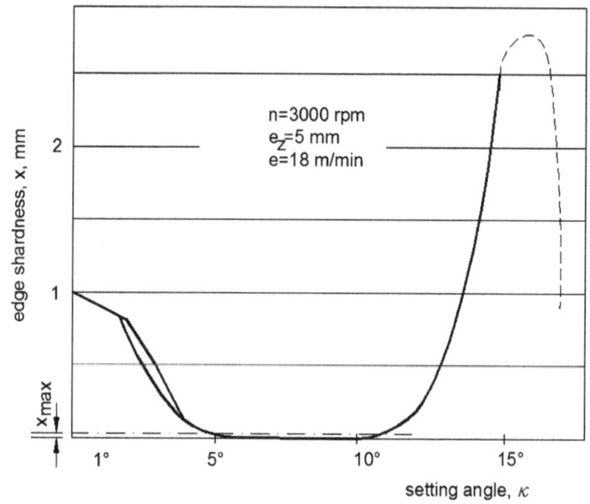

Fig. 3.21 Variation of edge splintering as a function of setting angle [4, 5]

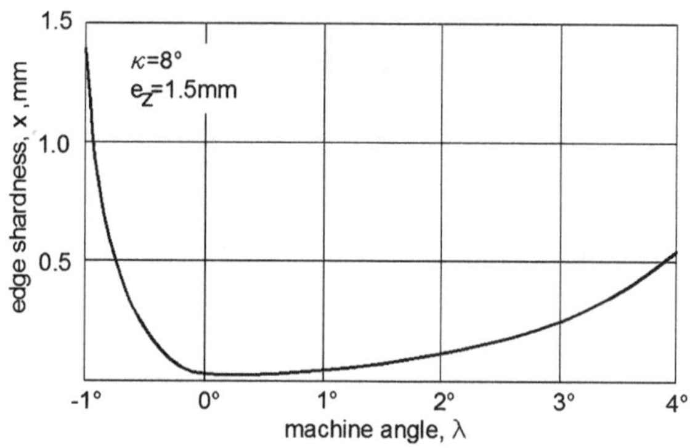

Fig. 3.22 Edge quality variation with tilting (machine) angles [4, 5]

and the chip thickness is

$$h = e_z \cdot \sin \kappa$$

Drills have a rotation and a feed motion, so their moving clearance angle decreases the true clearance angle. In the axis of a drill the moving clearance angle is 90°, so the central part of the drill always compresses the material. To reduce the compressed area as much as possible, the ratio of the tangential speed and feed rate (*v*/*e*) should be kept between 30 and 60.

This ratio is calculated from the following equation

$$\frac{v}{e} = \pi \frac{d}{e_n} \quad \text{and} \quad \tan \alpha' = \frac{e}{v}$$

where α' is the moving clearance angle.

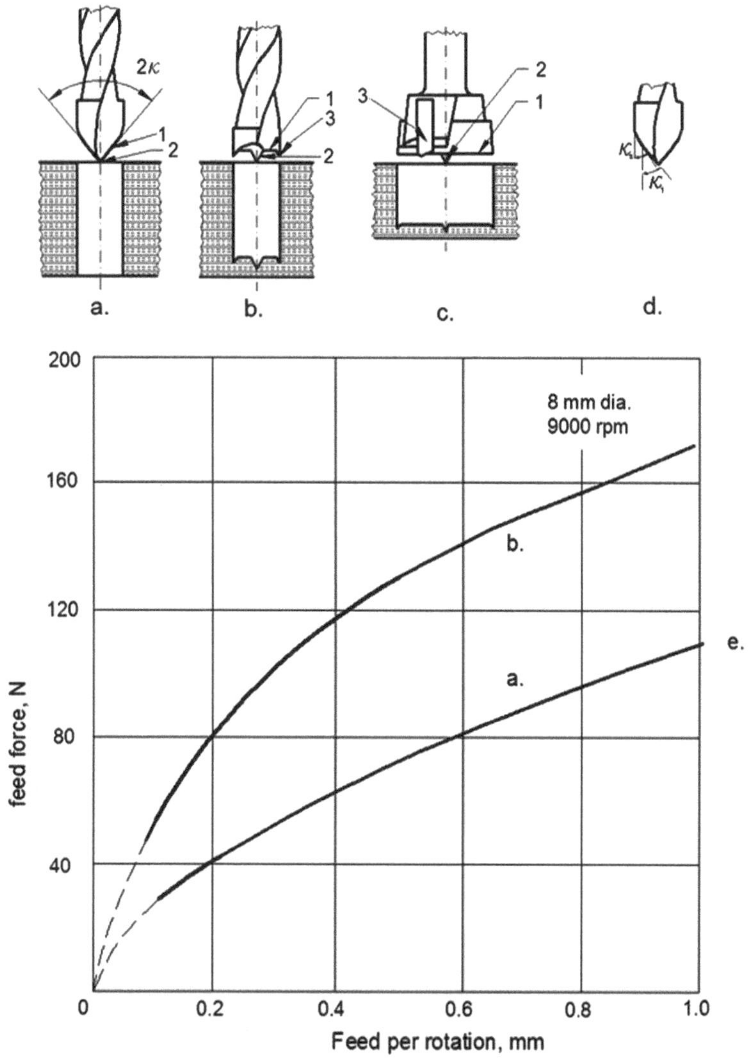

Fig. 3.23 Different machine bits using for through and dead-end drilling (**a-d**); feed force vs. feed per rotation (**e**) for drill types **a** and **b**. 1—cutting edge; 2—centre drill; 3—spur edge [7]

The rotation speed of drilling is generally 4,000–9,000 rpm but higher rotation speeds up to 20,000 rpm can be used.

Higher rotation speeds create some problems drilling dead-end holes, because the deceleration of the feed is limited to 3–4 m/s^2 [7].

The surface roughness of the hole depends on the feed per revolution. Drill types b and c are less sensitive to higher feeds per revolution because they have spur edges. These edges generally have a negative rake angle ensuring a better surface quality.

Special attention should be paid to the quality of the edge of the hole. Boring through holes, edge breaks and deformed edges (burr formation) often occur. The bottom edge is especially sensitive to edge breaking. In many cases, the only method to avoid burr formation is to use scrap wood under the worked material when drilling through holes. This is especially true for particleboard, MDF and covered coated boards.

The edge quality depends on the sharpness of the drill and also on the cone angle. A smaller cone angle gives a better quality edge of the hole. To do this, the cutting edge of the twist drill can be designed with two different cone angles, as shown in Fig. 3.23d. The second major influencing factor is the feed per revolution. A higher feed per revolution may be used for hard woods (beech) than for softwoods.

The central part of the drill, generally within the 0.1 R circle area, compresses the worked material. This is associated with a relatively high feed force (Fig. 3.23e). As a rough estimate, the feed force has the same magnitude as the cutting force reduced to the outer radius of the drill.

Due to the small size of drilling tools, the energy requirement of drilling is generally low. Measuring the energy consumption it seems to be advisable to measure the driving torque and not the cutting force.

The power consumption is:

$$P = \frac{M \cdot n}{9.55} = M \cdot \omega$$

where n means the rotation speed, rpm and ω is the angular speed, rad/s. The driving torque can be calculated using Eq. (3.8a) multiplied by the drill radius (the cutting force is related to the outer radius of the tool) in the following way:

$$M = (A + B \cdot h) \cdot b \cdot r = (A + B \cdot h)\frac{r^2 \cdot z}{\sin \kappa}$$

with $b = z\frac{r}{\sin \kappa}$ and $h = e_z \sin \kappa$

where the angle κ is the half of the drill cone angle, z is the number of edges. Using twist drills with cone angles between 60 and 80°, the constants in the equation have average values of $A = 34$ 1/cm and $B = 1250$ 1/cm^2. In this case the torque is obtained in Ncm.

Drills with spur edges (used to produce blind holes) generally have lower torque requirement than twist drills. Using high speed, however, the power requirement may be around 1 kW (diameter 8 mm and $n = 10{,}000$ rpm).

The cutting energy required for drilling related to the unit volume of solid wood is in the same magnitude as at the common woodworking operations: 17–25 kWh/m^3; (see Table 4.1).

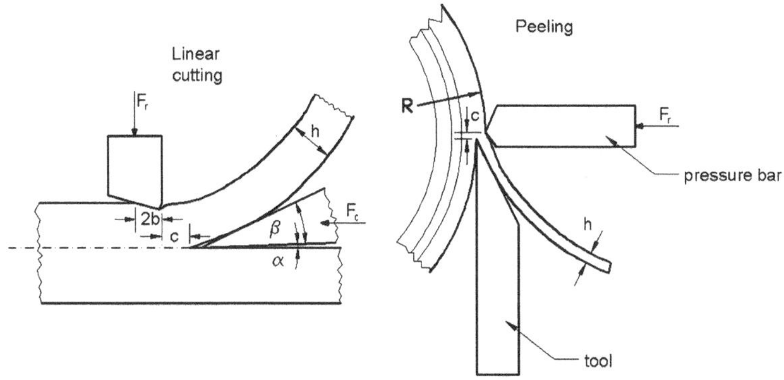

Fig. 3.24 Veneer production with linear cutting and peeling

3.6 Veneer Cutting and Peeling

The two main procedures for veneer production are linear and rotary cutting (peeling). The difference between the two cutting processes can be seen in Fig. 3.24. Both use a small bevel angle around 20 degrees; the clearance angle is as small as possible, generally between 0.5 and 1.5 degrees.

The pressure bar is important in the production of quality veneer. To produce good quality veneer, the nosebar position, both vertically and horizontally, must be optimally set over the knife. The setting parameters are the *relative deformation* of the veneer (z_0/h) and the *horizontal gap* (lead), which is the distance c between the rear edge of the nosebar and the cutting edge related to the veneer thickness (c/h).

The relative deformation of the veneer under the nosebar can be achieved by a prescribed setting of the nosebar relative to the cutting edge. The prescribed deformation of the veneer z_0 is 10–20 % of the veneer thickness. Another possible method may be to pre-stress the nosebar with a given constant load (e.g. spring load). Practical loads for nosebars are 200–300 N/cm.

Since the actual radial force on the nosebar is not constant, it varies in relation to its average value. That is due to inhomogeneities of the wood material and to oscillations of the tool. The force amplitude variation related to the average value is between 15 and 30 %.

The variation of nosebar radial forces as a function of the relative deformation of veneer 2 mm thick is shown in Fig. 3.25, [8]. As the relative deformation varies, the contact surface of the nosebar does not remain constant; therefore, the radial force increases more than linearly. The horizontal gap of the nosebar also influences the radial force. With a small horizontal gap, the rake face of the tool hinders the free deformation of the wood so the radial force on the nosebar will slightly be higher.

A generalization of the radial nosebar force was found as a reduced force on the vertical axis $F_r/h^{1.5}$ (Fig. 3.26). For very small deformations, especially for

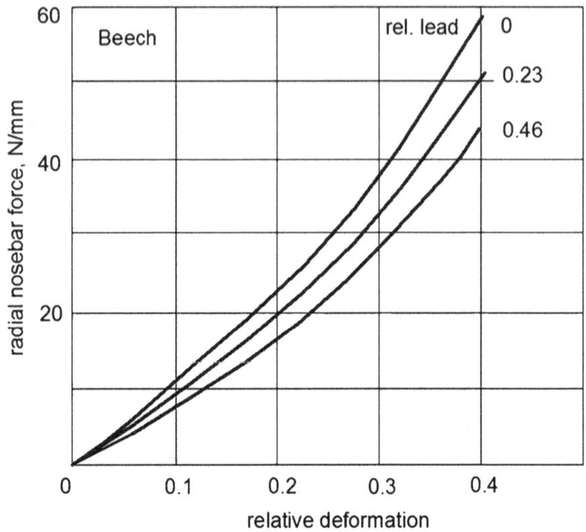

Fig. 3.25 Radial nosebar force depending on relative deformation and relative lead [8]

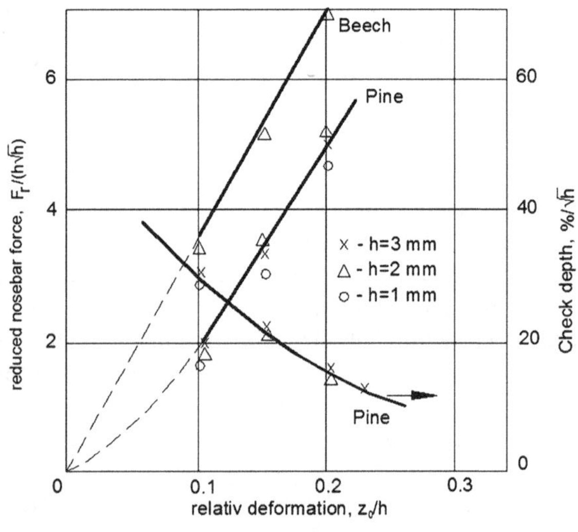

Fig. 3.26 Generalized nosebar forces and check depth as a function of relative deformation [8]

softwoods, some deviation from the straight line is obtained due to the surface roughness causing a smaller modulus of elasticity in the thin surface layer. This effect is smaller for hard woods.

The nosebar slides on the wood surface causing friction forces. These forces can be roughly estimated as true friction forces using friction coefficients between 0.32 and 0.38. The friction forces will be added to the cutting force.

The tangential cutting forces depend on the species of wood and cutting parameters. The main cutting parameters are the veneer thickness, the relative deformation of veneer and the horizontal gap (relative lead). The temperature of

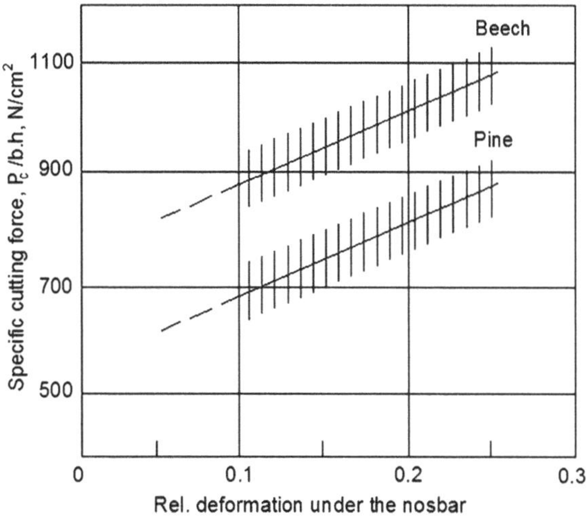

Fig. 3.27 Specific cutting forces for peeling pine and beech, cutting thickness between 1 and 3 mm [8, 9]

the wood and the cutting speed has a smaller influence. The specific cutting force related to the thickness *h* (*h* is given in cm) can be approximated with a straight line as a function of the relative deformation of the nosebar. Measurement results for beech and fir are presented in Fig. 3.27 for veneer with thicknesses between 1 and 3 mm [8, 9].

Excessive relative deformation may result in a rough surface and a large power requirement.

The quality of veneer can be characterised by the surface roughness and the relative checking depth of a veneer surface in tension. The checks depend on the veneer thickness, wood species, temperature of the wood and the nosebar pressure.

The nosebar pressure or relative deformation under the nosebar is a powerful means to reduce surface checking. Check depth is generally represented as a relative value related to the veneer thickness in percent. If this value is divided by the square-root of the veneer thickness, then this reduced check depth is more or less independent of the veneer thickness. This relationship for pine wood is given in Fig. 3.26.

3.7 Sanding

Abrasive sanding is one of the most important steps to produce high quality wood products. Sanding is an abrasive cutting process in which material is removed by an unusual cutting edge with a negative rake angle (Fig. 3.28). The cutting tool is a hard sphere with a radius *R* embedded in a carrying material (sand paper). Chip formation depends on the compression yield stress of the wood, the vertical load on the cutting edge, the radius of the cutting sphere (grit size) and the cutting speed.

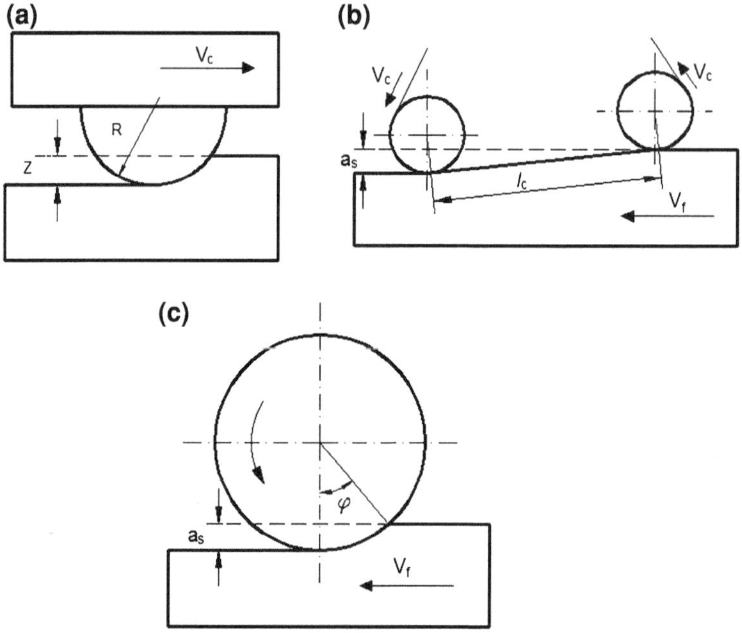

Fig. 3.28 The working principle of a spherical cutting tool (**a**), linear sanding (**b**) and cylindrical sanding (**c**)

The grit size is determined by the number of meshes per inch of the sieves used for screening. The real size of the grit is generally normal distributed between two successive sieves. In practice, the average grit size is used. Huge amounts of grit are embedded in the carrying base material with overlapping paths.

The main purpose of sanding is to achieve the best possible surface quality characterized by the different s*urface roughness parameters* (see in Sect. 8.1). The most important operational parameters are the surface pressure, the feed speed, the grit size and the sanding speed. The previous machining also has a considerable influence on the operational parameters: the thickness of the layer to be removed strongly depends on the previous machining quality.

The abrasive material (grit) is generally made of either aluminium oxide (Al_2O_3) or silicon carbide (SiC). They have different physical properties in their hardness and heat conduction coefficient. Silicon carbide has a hardness of 9.7 (Mohs) and $\lambda = 41$ W/m.K, while aluminium oxide has 9.0 and $\lambda = 20$ W/m.K. The high heat conduction coefficient is very important for sanding wood composites containing adhesives. In practice, corundum is mainly used for sanding solid woods but, silicon carbide gives better results on particle board and MDF. At high temperatures, the adhesive may soften causing smearing and clogging. Choosing an abrasive with a better heat conduction coefficient may prevent such difficulties.

Sanding is a special cutting process characterized by a negative rake angle of the cutting edge and by the random position of grit embedded in the holding tissue.

An accurate description of the sanding process is not possible; nevertheless, some general relationships can be derived. The deformation of a spherical rigid indenter acting on a plane surface, is given by the following equation [10]:

$$z = const \cdot \left(\frac{F}{E}\right)^{\frac{2}{3}} \cdot \frac{1}{R^{\frac{1}{3}}}$$

where F is the force, R is the radius of the indenter and E is the modulus of elasticity.

Moving the spherical indenter horizontally, the chip cross section is an arc of a circle with a height z and its value can be approximated with the following simple equation:

$$A = \frac{\pi}{2} R^{0.55} \cdot z^{1.45}$$

Using these two equations, and assuming that grit always has the same relative cover of the surface, the chip cross section assumes the following form [11]:

$$A \cong const \cdot \frac{F}{E} = const \cdot p$$

where p means the contact pressure on the sanding belt or drum. If the workpiece is stationary, then the amount of material removed in relation to the unit width is proportional to the length of sanding path

$$\frac{V}{b} = const \cdot Av_c t = const \cdot pv_c t$$

where v_c is the cutting speed, and t is the sanding time. When the sanding surface area is constant, the thickness of the material removed also varies linearly as a function of time.

It is advisable to measure the weight instead of volume and so we obtain a dimensionally correct equation in the form:

$$\frac{V\gamma}{b} = B \cdot Lpv_c t \quad [\mathrm{N/cm}] \tag{3.11}$$

where B is a material constant related to the unit length, L is the length of the workpiece ($L \leq L_c$), and γ is the specific weight of the wood, $\mathrm{N/cm^3}$. For beech wood, the material constant B has values between $4.6*10^{-8}$ and $5.4*10^{-8}$ 1/cm.

The thickness of the removed material is also time dependent:

$$a_s = \frac{V}{b \cdot l} = B \cdot \frac{p \cdot v_c \cdot t}{\gamma}$$

However, the workpiece generally moves under the sanding belt or drum with a feed speed v_f. The sanding time for a given point depends on the contact length L_c and feed speed

Table 3.1

Wood species	Beech	Alder	Oak	Pine
k, g/cm^{2}min	0.25–0.3	0.4–0.5	0.2–0.25	0.3–0.4

Note: the above Figures are valid for $v_c = 8$–9 m/s

$$t = \frac{L_c}{v_f}$$

where L_c is the maximum contact length of the sanding device (see Fig. 3.28). The removed material is now calculated as follows:

$$\frac{V \cdot \gamma}{b} = \frac{B \cdot L_c^2 \cdot p \cdot v_c}{v_f} \tag{3.11a}$$

The thickness of the material removed is calculated as

$$a_s = \frac{V}{b \cdot L_c} = B \cdot L_c \frac{p \cdot v_c}{\gamma \cdot v_f} \tag{3.11b}$$

The amount of wood removed related to the unit area and unit time is given by

$$k = \rho \frac{a_s}{L_c} v_f \quad \left[g/cm^2 \min\right] \tag{3.12}$$

where ρ means the density of the wood. Approximate values of the amount of wood removed for a surface pressure of 1 N/cm^2 is given in Table 3.1.

The amount of wood removed by sanding belts continuously decreases as a function of working time due to the wear process. Wood species containing resins work less effectively in the first 20–30 min.

Keeping in mind Eq. (3.11b), the following dimensionless number can be derived:

$$\Pi = \frac{p \cdot v_c}{a_s \cdot \gamma \cdot v_f}$$

which should be invariant for sanding a given wood material. Combining Eqs. (3.11b) and (3.12), the amount of wood removal k can be related to the material constant B as

$$k = B \cdot \rho \cdot \frac{p}{\gamma} \cdot v_c \tag{3.12a}$$

or

$$B = \frac{k \cdot \gamma}{\rho \cdot p \cdot v_c} \tag{3.12b}$$

or using Eq. (3.12), we get

$$B = a_s \frac{\gamma}{p \cdot L_c} \frac{v_f}{v_c} \tag{3.12c}$$

In the possession of some experimental results, e.g. thickness of the removed material a_S, all important operational parameters can easily be calculated for arbitrary initial conditions. The continuous wear process should be taken into account resulting in a decrease of all performance parameters as a function of working time. The simplest way to account for the wear process is to use an exponential function in the form:

$$B = B_0 \cdot e^{-\beta \cdot t}$$

where B_0 is the initial value of the material constant B and the constant β characterizes the rate of decrease as a function of time. Here the wood removal is time dependent and its value is obtained by integration over the required time span t:

$$\frac{V}{b} = B_0 \cdot L_c \cdot v_c \frac{p}{\gamma} \int_0^t e^{-\beta \cdot t} dt$$

which yields

$$\frac{V}{b} = B_0 \cdot L_c \cdot v_c \cdot \frac{p}{\gamma} \frac{1}{\beta} \left(1 - e^{-\beta \cdot t}\right) \tag{3.13}$$

The value of β is dependent on the sanding belt speed. As a first approximation, β may be taken as proportional to the cutting speed v_c in the form:

$$\beta = D \cdot v_c$$

There is an average value of $D = 2.2*10^{-7}$ 1/cm from experimental data on processing beech wood. Using a sand belt speed of 20 m/s = 120.000 cm/min, the value of exponent β will be 0.0264 1/min or 1.584 1/h. The material constant B_O, is between $4.6*10^{-8}$ and $5.4*10^{-8}$ 1/cm.

Keeping in mind the above expression for β, the material removal related to the unit width can be given by the following general equation:

$$\frac{V}{b} = \left(\frac{V}{b}\right)_{\max} \cdot \left(1 - e^{-\beta \cdot t}\right) \quad \text{cm}^3/\text{cm} \tag{3.13a}$$

With

$$\left(\frac{V}{b}\right)_{\max} = \frac{B_0 \cdot p}{D \cdot \gamma} \cdot L_c$$

$(V/b)_{max}$ means the maximum theoretical stock removal using the belt for an infinitely long time. In practical cases, the actual stock removal is 75–80 % of the

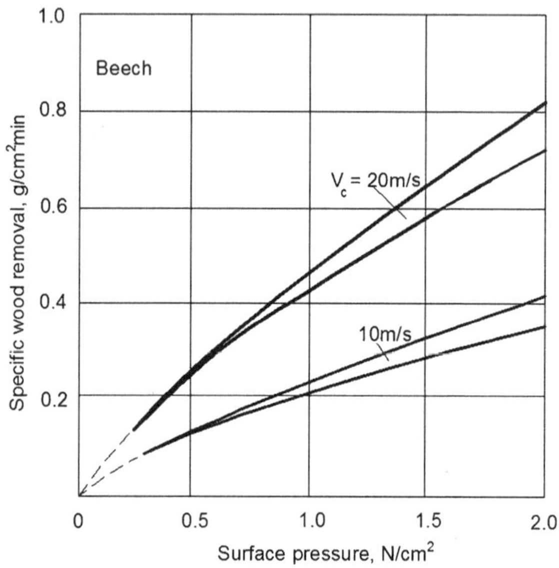

Fig. 3.29 Wood removal as a function of surface pressure. Grit size is P80 and P120

maximum value. The working time required for this depends on the sanding belt speed: for $v_c = 20$ m/s the working time is approximately 1.0 *h* and for $v_c = 10$ m/s the time is doubled.

Exponent β probably depends on the contact pressure. Experiments are needed to clarify this relationship in detail.

Limited experimental results show that the theoretically obtained linearity between stock removal and surface pressure in a wider range does not hold. When the surface pressure increases, more chips are in the contact surface and they may hinder the full penetration of grit into the wood surface corresponding to the given pressure. Therefore, slightly less wood is removed at higher pressures than expected from the linear law. An appropriate correction can be made by modifying the maximum stock removal using an exponent for p less than one:

$$\left(\frac{V}{b}\right)_{\max} = \frac{B_0}{D} L_c \frac{p^{0.8}}{\gamma}$$

Then the material constant B_0 has to be modified and it is between $3.5*10^{-8}$ and $4.1*10^{-8}$ 1/cm (beech wood).

The same thing must be done to calculate the effect of grit size on stock removal The free height of fine grit is very low and, therefore, the chips on the sliding surface hinder the penetration of their edges into the wood surface. Experiments show that the wood removal, using P 180 or finer grits, decreases considerably [12]. Figure 3.29 shows the wood removal as a function of surface pressure for different grit sizes. The relationship is theoretically linear; in practice, a slight deviation is obtained in the higher surface pressure range. Wood removal is inversely proportional to the feed speed due to the decreasing time in action on a

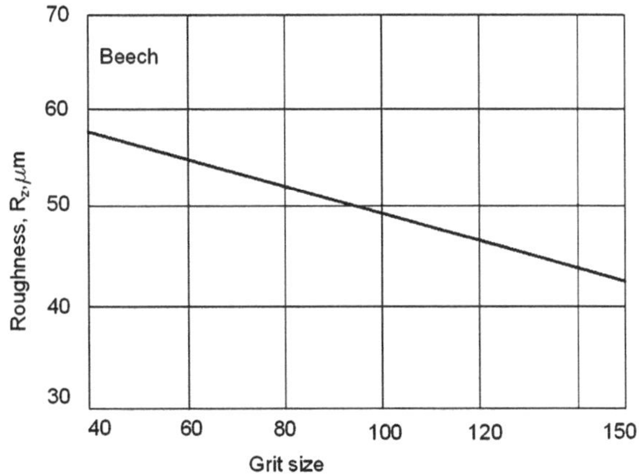

Fig. 3.30 Effect of grit size on the roughness parameter R_z for beech wood. Sanding speed 20 m/s [13]

given surface. Therefore, a strong correlation exists between thickness of the layer to be removed and the feed speed. The sanding speed and the grit size of the belt are the most important operational parameters to obtain a good surface roughness. Figure 3.30 demonstrates the effect of grit size on the surface roughness parameter R_z for beech wood. The relationship is practically linear in the given grit size range [13]. An increased sanding speed reduces roughness, but it is less effective compared to the grit size. An increased sanding speed from 10 to 25 m/s improves the roughness values by $\Delta R_z = 3$ to 4 µm.

References

1. Sitkei, G. et. al.: Theorie des Spanens von Holz. Fortschrittbericht No.1. Acta Fac. Ligniensis, Sopron (1990)
2. Sitkei, G., Horváth, M.: Log dynamics during sawing in frame saws. In: Proceedings of 12th IWMS Kyoto, pp. 327–334 (1995)
3. Westkämper, E., Freytag, J.: PKD-Schneidstoff zum Sägen melaminharzbeschichteter Spanplatten. IDR 1/1991, S. 46–49
4. Lang, M.: Untersuchungen zur Entwicklung eines kombinierten Plan- und Umfangsfräsverfahrens mit kegelstumpförmigen Werkzeugen. PhD Dissertation, Dresden–Sopron (1989)
5. Tröger, J., Lang, M.: Hobelfräsen mit vermindertem Wellenschlag. HOB, 11/1990, S. 43–49
6. Кряжев, Н. А.: Цилиндрическое и коническое фрезерование древесины. (Cylindric and cone-shaped milling tools for wood machining) Лесная Пром., Москва (1963)
7. Westkämper, E., Kisselbach, A.: Leistungsteigerung beim Bohren. HOB 1995, S. 94–103
8. Mothe, F., Marchal, R.: Influence of nosebar settings on tool instabilities in the peeling process. In: Proceedings of 15th IWMS Los Angeles, pp. 309–328 (2001)
9. Bakar, E., Marchal, R.: Efficiency of automatic adjustment of a slanted-elastic pressure nosebar in peeling. In: Proceedings of 16th IWMS Matsue, pp. 815–821 (2003)
10. Timoshenko, S., Goodier, J.: Theory of Elasticity. Mc. Graw-Hill, NewYork, (1951)
11. Scholz, F., Ratmasingam, J.: Optimization of sanding process. Proceedings of 17th IWMS Rosenheim, pp. 422–429 (2005)

12. Dobrindt, P.: Optimiertes Schleifen von MDF—und Spanplatten. Workshop Tagungsband, 8.Holztechn. Koll. Braunschweig, S. 125–136 (1991)
13. Siklinka, M., Ockajova, A.: The study of selected parameters in wood sanding. Proceedings of 15th Int. Wood Machining Seminar Los Angeles, pp. 485–490 (2001)

Chapter 4
The Energy Requirement of Cutting

The energy consumption of woodworking operations is significant part of the total production cost. Therefore, it is important to select proper tools with optimum operational parameters.

After a short theoretical consideration, the energy requirements of different woodworking machines are treated. Generally, working diagrams are constructed to facilitate the consideration of several variables influencing the specific force or energy requirements.

In many cases other important features such as production rate, surface quality and tool maintenance cycle must be considered to minimize manufacturing costs. All operating parameters must be analysed to determine the best specifications for each production line.

4.1 Theoretical Considerations

In the discussion of the theory of cutting we derived and proved that the cutting force per unit length can be expressed with the simple linear equation [1].

$$\frac{P_h}{b} = A + B \cdot h$$

The A and B constants in the equation depend on the mechanical properties of the wood and the cutting angle δ of the tool. The δ angle primarily influences the value of the B constant and its effect can be given with the following empirical equation (see Fig. 1.8):

$$B = B_0 + a(\delta - 40)^n$$

where

B_0 is the minimum value of constant B at $\delta = 40^\circ$.

E. Csanády and E. Magoss, *Mechanics of Wood Machining*,
DOI: 10.1007/978-3-642-29955-1_4, © Springer-Verlag Berlin Heidelberg 2013

Processing the experimental data gave the following values to the constants:

for beech $a = 1.5 \times 10^{-3}$ N/mm^2 and $n = 2.75$;
for pine $a = 5.5 \times 10^{-4}$ N/mm^2 and $n = 3$.

Some experimental results have indicated that for small chip thickness ($h = 0.1$–0.2 mm) the optimum cutting angle shifts to slightly higher values falling in the range of 45–55 degree [3].

Certain tools, especially saws, have some special features. Previously we have not dealt with the role of the secondary cutting edge and we have referred the cutting force per unit length to the length of the principal cutting edge. In the most cases this does not cause error.

The main cutting edge of saws is relatively short and the length of the secondary cutting edge in relation to the length of the main cutting edge (15–30 %) is not negligible. The work of the secondary cutting edges of a saw tooth is shown in Fig. 4.1. Shearing occurs on two sides of the tooth at the deformation zone of the cutting (F cross section). The width of the sheared cross section depends on the rupture strain. Its value has not yet been exactly determined. But it is probable that the sheared cross section is proportional with chip thickness h, so the value of the shearing force is:

$$P''_h = 2 \cdot c \cdot h \cdot \tau$$

where:

c is a constant ($c \cong 0.3$) and
τ means the shear strength in the direction of the cutting

This shearing force depends on h chip thickness, so it will increase the B constant in the general cutting force equation. This explains why the cutting force diagrams of the saws are steeper as a function of chip thickness than those for milling and planing machines [2, 3].

An increase of the cutting height generally increases the cutting force. The reason is that the mass of chips in the gullet rubs against the wall of the gap. The more chips are in the gullet the greater the friction force will be. The development of side pressure in the gullet was already demonstrated in Fig. 3.2. The value of the additional friction force on the wall of the gap can be calculated as

$$S = \mu \cdot 2 \cdot F \cdot \sigma_h$$

where:

F the cross section of the gullet,
μ the friction coefficient,
σ_h the side pressure of the chips in the gullet.

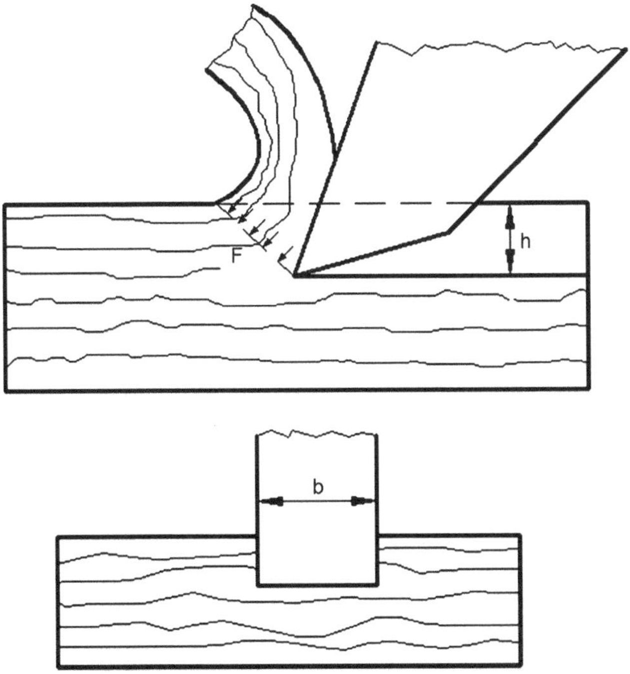

Fig. 4.1 The work of the secondary cutting edge of a saw tooth

The volume of the chips can be decreased by 20–30 % with relatively small pressure. But a greater compaction in the gullet substantially increases the side pressure and the friction force.

The amount of chips in the gullet can be expressed with the *gullet feed index*:

$$\varphi = \frac{V_{chip}}{V_g} = \varepsilon \frac{V_{sw}}{V_g}$$

where:

V_{sw} volume of solid wood,
$V_{chip.}$ volume of chips.

The φ gullet feed index has the unit value if loose chips fill the gullet. If the chips are totally compacted, then $\varphi = \varepsilon$ ($\varepsilon = 3$–4).

The power requirement can be generally written as follows:

$$N = N_0 + P \cdot v$$

where:

N_0 the idling power,
P all of the forces,
v the speed.

The force is approximately proportional to the chip thickness or to the feed per tooth at cutting, while the number of teeth in action is:

$$z' = \frac{H_\Sigma}{t}$$

where:

H_Σ sum of the cutting heights,
t the tooth-pitch.

Keeping in mind Eq. (3.1), the power requirement can be written as follows:

$$P \cdot v = const \cdot e_z \cdot v = const \cdot e \cdot H_\Sigma$$

And with this the total power requirement is given by the following general equation:

$$N = N_0 + const \cdot e \cdot H_\Sigma$$

This equation shows that it is advisable to use the variable $e \cdot H_\Sigma$ in processing experimental results.

4.2 The Energy Requirements of Bandsaw

The bandsaw is a generally used machine in the wood industry for cutting, profile cutting, ripsawing and splitting logs. Equation (3.3), in which most of the variables in a given case are a constant can be written shortly as follows:

$$e_{\max} = \frac{const}{H}$$

This means that the maximum feed speed is determined by the height of the sawn cross section.

Cutting of a given cross section H the gullet feed index increases linearly and it reaches its maximum value at the exit from the wood. In the same situations, the value of $\varphi_{\max}$ is larger as the height of the sawn cross section increases. The cutting resistance of Scotch pine is demonstrated in Fig. 4.2 depending on the feed per tooth at different cutting heights. In the figure the constant lines of φ_{max} are included and as a comparison to the cutting forces, the cutting resistance of a milling machine with a similar cutting angle is also given.

The equation of the curves can be given in the following fashion:

$$\frac{P_h}{b} = 9 + 43 \cdot e_z + 7 \cdot \sqrt{e_z} \cdot \varphi_{\max}^{0.9} \tag{4.1}$$

In the above equation, a new term is added to take into consideration the friction forces arising on the side walls.

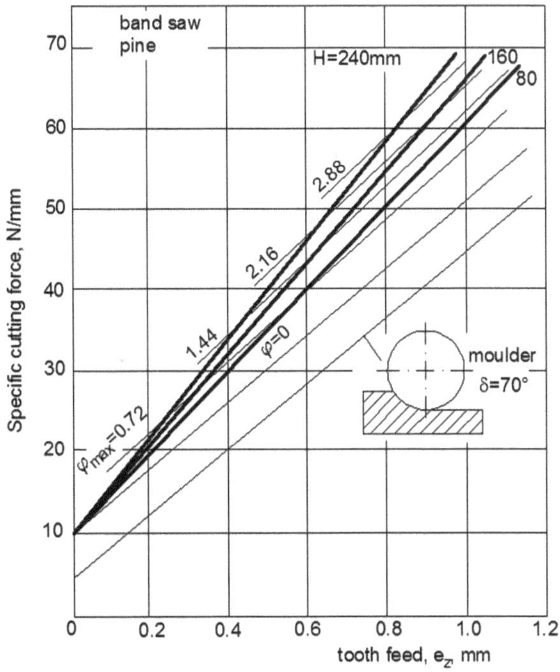

Fig. 4.2 The cutting force of bandsaws depending on the feed per tooth for different cutting heights [1]

Figure 4.3 shows the power requirement of cutting depending on the sawn cross section in the unit time. The power requirement is a little bit smaller for a thicker chip. That can be explained by the effect of the first term *(A)* in the cutting force Eq. (1.8a) which will be smaller with a larger thickness of the chip. The equation of the curves in the figure can be given as follows:

$$N = 146 \cdot (e \cdot H)^{1.15} + 160 \cdot (e \cdot H) e^{\frac{-e_z}{0.65}} \quad (4.2)$$

The sharpness of the tool substantially influences their cutting force. The data refers to sharp tools, when the radius of the edge is between $\rho = 20$ and 30 μm; (see Sect. 7.1).

4.3 The Energy Requirements of Frame-Saw

Frame-saw have been the base machine of the primary wood processing for more than 100 years. The frame-saw has some disadvantages in spite of its mature technical design. Its moving masses can be balanced only in the first order. Therefore its speed is only 320–340 rpm today and its average cutting speed is $v_{av} = 6-7$ m/s.

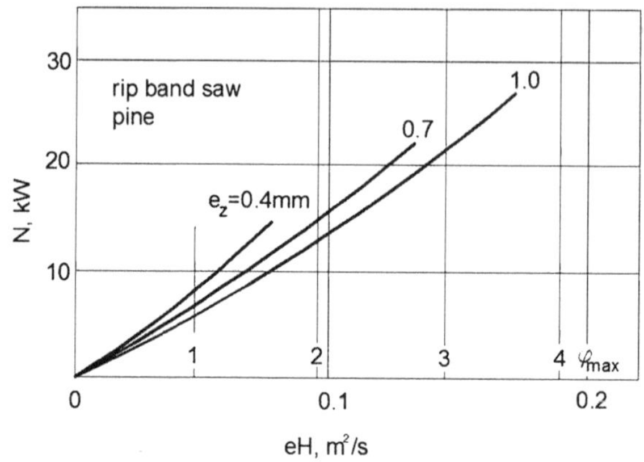

Fig. 4.3 Power consumption of the bandsaw depending on the sawn cross section in the unit time [1]

The most important factor is the height of the cutting *H*. The pitch is t = 25–30 mm; the cross section of the gullet is 2.3–2.8 cm^2, so the maximum feed speed can be given approximately as follows:

with spring set teeth

$$e_{\max} \cong \frac{0.95}{H}$$

with swage set teeth

$$e_{\max} \cong \frac{0.62}{H}$$

where the *H* cut height must be substituted in m.

The idling power of a frame-saw depends on the driving method of the auxiliary equipment. Modern machines are provided with hydraulic devices, and with these we can calculate with $N_0 = 20$ kW value. The cutting power requirement for black locust and Scotch pine are shown in Fig. 4.4. The power requirement of a frame saw for pine is given by:

$$N = 20 + 4.1 \cdot e \cdot H_\Sigma$$

for black locust:

$$N = 20 + 6 \cdot e \cdot H_\Sigma$$

where $e \cdot H_\Sigma$ must be substituted in m^2/min.

In order to calculate the total cutting height, the sawing pattern (cutting distance) of the log is needed (Fig. 4.5).

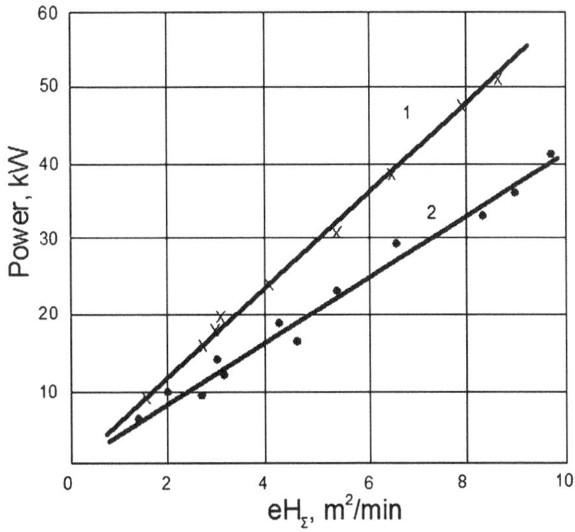

Fig. 4.4 The power consumption of a frame saw depending on the total sawn cross section. *1* black locust, *2* pine [1]

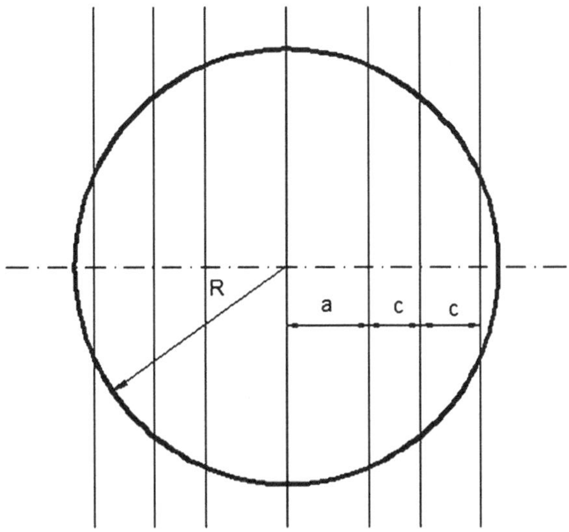

Fig. 4.5 Cutting pattern of frame saws [1]

The total cutting height can be calculated as

$$H_\Sigma = 4 \cdot \left[\frac{R}{2} + \sum_{i=0}^{k} \sqrt{R^2 - (a + i \cdot c)^2} \right]$$

where $R \geq a + i \cdot c$

and i is the number of cuts to the right of the centre line.

The theoretical sawing capacity of a frame saw is influenced by the diameter of the log and the feed speed

$$V = \frac{d^2\pi}{4} \cdot e \cdot 60 \quad (\mathrm{m}^3/\mathrm{h})$$

Keeping in mind that the feed speed decreases with increasing log diameter, therefore, the sawing capacity is proportional to the log diameter. As a rough estimate, the sawing capacity is given by the following equation

$$V = \eta_t \cdot 45 \cdot d \quad (\mathrm{m}^3/\mathrm{h})$$

where the log diameter d should be substituted in m, and η_t is the time exploitation coefficient.

A working diagram is needed to facilitate the selection of the optimum operational parameters (Fig. 4.6) [1]. The starting point is the log diameter and from this we obtain the maximum feed speed and the cut surface in the unit time. Furthermore, in the lower part of the diagram we find the power requirements for different species of wood.

The power to cut one cubic meter of wood depends on the log diameter and increases quickly under $d = 25$ cm. We can usually calculate 4–5 kWh/m^3 specific energy input [1].

4.4 The Energy Requirements for Planing and Milling

The planer and the milling machine are the classic rotating tools in the wood industry. We can use them in relation to the direction of the feed in climb or counter cutting run; (see Fig. 3.15). The magnitude of the peripheral force is practically independent of the run, but climb cutting usually results in higher surface quality.

The change of the specific cutting force is shown in Fig. 4.7 depending on the direction of the wood grain and the cutting angles δ [2]. We get straight lines as a function of chip thickness according to the cutting theory. The increase of the angle δ increases the cutting force, but generally it improves the surface quality. This can be explained by the increase of compressive stresses in the surroundings of the edge.

Power consumption is demonstrated in Fig. 4.8 depending on the sawn cross section per unit time [1]. From the diagram we can read the effect of cutting depth (H) and the feed per tooth.

The power requirement can be calculated from the following equation:

$$N = \psi \cdot R(A \cdot b + B \cdot b \cdot e_z)\frac{n}{9.55} \cdot \frac{z \cdot \varphi}{360} \tag{4.3}$$

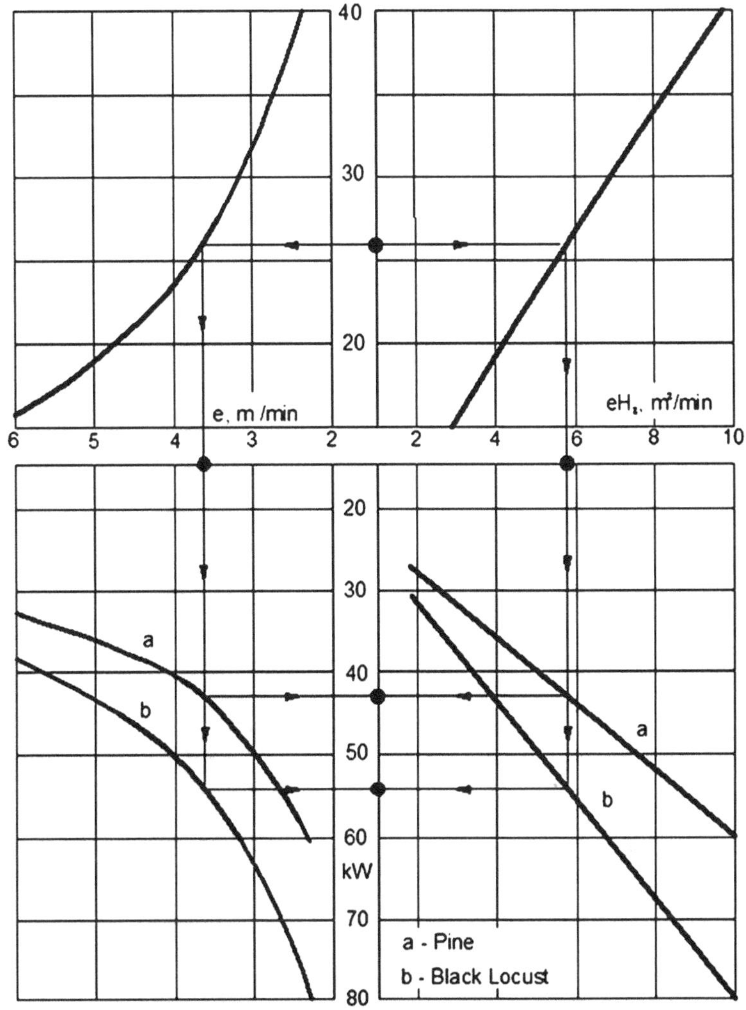

Fig. 4.6 Working diagram for frame-saws [1]

where:

R the radius of the tool,
ψ the correction factor (see Fig. 2.2b),
φ angle of cutting arc,
z tooth number,
n rotation speed, rpm.

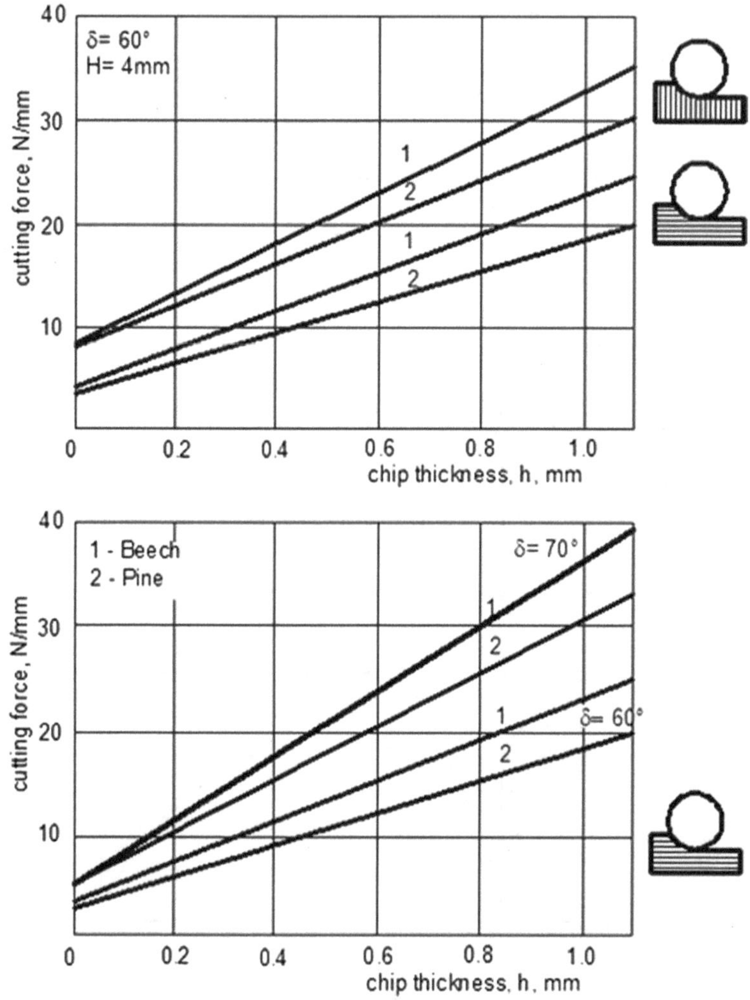

Fig. 4.7 The change of the cutting force depending on the cutting direction and cutting angle δ, for beech (*1*) and pine (*2*) [2]

4.5 Energy Requirements of Circular Saws

The working conditions of the circular saw is slightly similar to the band saw, but due to its cycloid movement, the cross section of the chip is variable; (see Fig. 3.10).

Forces acting on the main cutting edge are similar to forces acting on milling tools. In addition, forces acting on the secondary cutting edges and chip friction on the gap wall should be taken into account (see in Sect. 4.1).

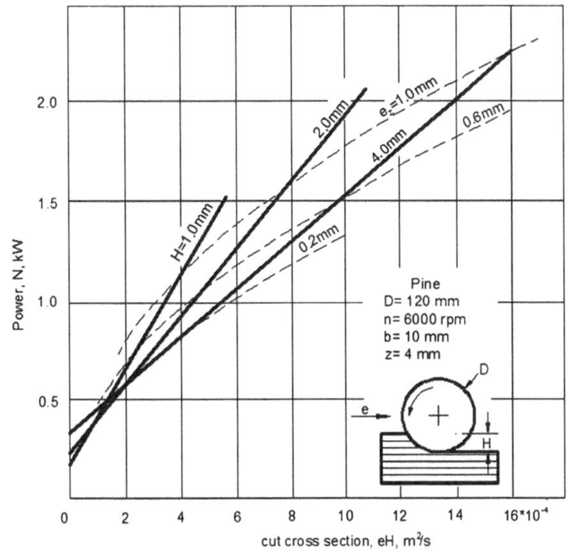

Fig. 4.8 The power consumption of milling [1]

The resultant tangential force can be calculated as follows:

$$P_t = z'(A \cdot b + B \cdot b \cdot h_{av} + 2c \cdot h_{av}\tau) \cdot \left(\frac{\varphi \cdot R}{H}\right) \tag{4.4}$$

where:

h_{av} the average chip thickness,
z' the number of teeth in action,
τ shear strength.

For the sake of simplicity, we assumed that the frictional forces increase with the relative arc length of the cutting. The third term in Eq. (4.4) increases the cutting force as a function of chip thicknesses as outlined in Sect. 4.1.

Using Eq. (4.4), the power consumption will be calculated as:

$$N = P_t \cdot \frac{v}{1000} \quad [\text{kW}]$$

The power consumption of cutting as a function of the cut cross section is illustrated in Fig. 4.9 for different cutting heights. We have drawn the constant e_z lines on the figure as well. We can see that the energy input (kWh/m^2) increases quickly as the cutting height increases and the feed per tooth decreases.

The energy consumption of circular sawing also depends on the cutting speed. Figure 4.10 shows experimental results using two different cutting speeds: (fir, cutting height 50 mm, $D = 400$ mm, $s = 2$ mm).

Higher cutting speeds require slightly more power for the same sawing performance due to higher cutting forces and ventilation drags.

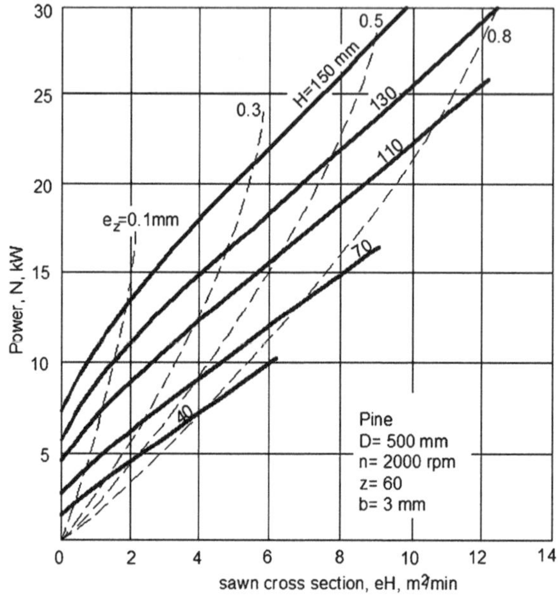

Fig. 4.9 The power consumption of a circular saw (cutting speed is 52 m/s) [1]

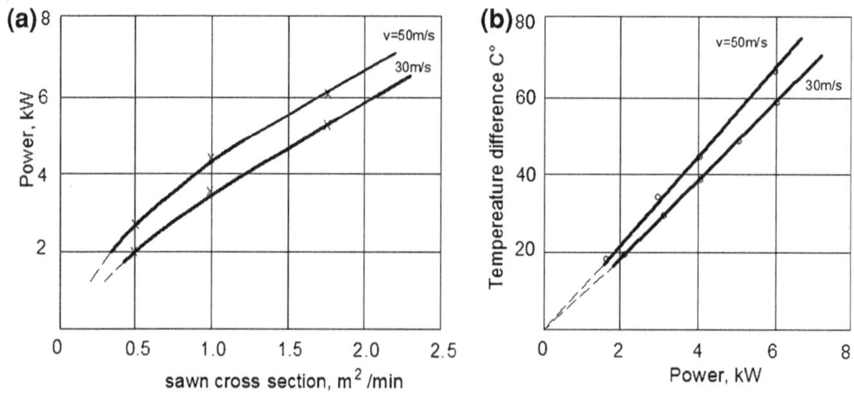

Fig. 4.10 Power requirement as a function of sawn cross section (**a**), and the relationship between the maximum temperature difference and power requirement (**b**) [3]

The friction forces associated with cutting will be turned into heat, warming up the tooth zone. Due to heat conduction and heat radiation to the air, a typical temperature profile will be developed and its shape can generally be described by a third power equation. This uneven heat distribution generates compressive tangential stresses in the saw blade reducing the quality of the sawing. The maximum temperature difference along the blade radius is generally proportional to the

Table 4.1

Machines	Pine wood	Beech wood
Bandsaw	20–22	25–28
Frame saw	20–22	25–28
Circular saw	18–20	24–27
Moulder	35–40	44–48
Disc sanding	70–100	80–120
Belt sanding	160–200	200–250

cutting power as demonstrated in Fig. 4.10b. The cutting speed has some influence on the maximum temperature difference.

Finally, an interesting comparison can be made concerning the energy of chip formation (kWh/m^3 solid wood) for different woodworking machines. Table 4.1 shows approximate average values for pine and beech woods.

Saws work with the same principle and their energy consumption differs very little. Moulding, milling and planing require more energy due to the mixed cutting directions. Sanding has the highest energy requirement, which may be from 5 to 10 times higher in comparison to saws. Especially belt sanders consume much energy due to friction forces on the belt and chip deformation between belt and workpiece.

Solid woods have a strong anisotropy and, therefore, the cutting force significantly differs in the main grain directions (see Fig. 4.7). Particle boards and fibre boards are more homogeneus materials requiring no distinction in cutting directions. At the same time, the type and quantity of gluing materials, the density of the boards may cause significant differences in the cutting force and energy requirements [4–6].

References

1. Sitkei, G., et al.: Theorie des Spanens von Holz. Fortschrittbericht No.1. Acta Fac. Ligniensis Sopron. (1990)
2. Saljé, E., Dubenkropp, D.: Zerspanbarkeitskennwerte beim Fräsen von Holz. Holz als Roh- und Werkstoff. S. 333–336 (1977)
3. Санев, В.: Обработка древесины круглими пилами. Изд. Лесная Пром. (Woodworking with circular saw) (1980)
4. Saljé, E., et al.: Erkenntnisse beim Fräsen von Spanplatten mit unterschiedlichen Schnittbedingungen. Holz als Roh- und Werkstoff. S. 501–506 (1985)
5. Saljé, E., Dubenkropp, D.: Das Kantenfräsen von Holzwerkstoffplatten. Holz- und Kunstoffverarbeitung. No. 4. (1983)
6. Saljé, E., Drückhammer J.: Qualitätskontrolle bei der Kantenbearbeitung. Holz als Roh- und Werkstoff. S. 187–192 (1984)

Chapter 5
Vibration of the Tools and Workpieces

Due to the elastic support and elastic nature of wood, the machine, the tool and the workpiece generally create vibration systems. The natural frequency and the deflection of the free vibration will be determined by the mass and the spring constant of the support. The more rigid the vibration system is, the smaller the deflections will be.

Wood working machines standing on a foundation form a vibration system. The vertical displacement of the machine table is usually 15–30 μm, and this does not significantly influence the accuracy of the woodworking and the surface quality. An exception is the frame-saw, which has remarkable vibration amplitudes due to the unbalanced masses of the sliding crank mechanism but a frame-saw is seldom used for doing precise work.

It is important to decrease the vibration of saws to achieve good sawing accuracy and surface quality. Saws are thin walled structural elements, and their bending stiffness is small. Therefore, significant lateral and torsional vibrations can develop.

One of the unique features of woodworking machines is that the workpiece moves during processing. The moving wood is pressed to the machine table by spring loaded rollers, and the wood vibrates against them.

In a CNC machine, the workpiece is fixed by pneumatic clamping devices which are not entirely rigid.

In the following, the vibration behaviour of the different woodworking machines will be treated in detail. Calculation methods and experimental results will be presented with practical recommendations.

5.1 Band Saw Vibration

The schematic layout of a band saw and the induced vibration patterns can be seen in Fig. 5.1. Lateral and torsional vibrations can occur inside the free span length, and these can appear to be superimposed.

E. Csanády and E. Magoss, *Mechanics of Wood Machining*,
DOI: 10.1007/978-3-642-29955-1_5, © Springer-Verlag Berlin Heidelberg 2013

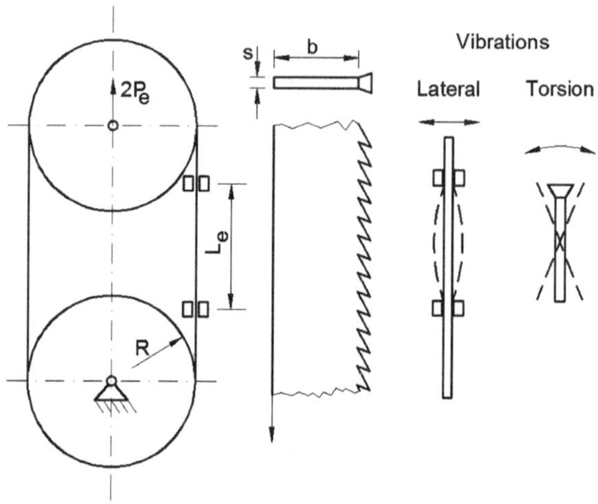

Fig. 5.1 The vibration forms of a band saw [1]

The natural frequency and the deflections are influenced by the following factors:

- pre-stressing force, P_e,
- free span length, L_e,
- the asymmetric heating of the band,
- external loads.

The vibrations are always created by the effect of excitation. Excitation can originate from the cutting forces, the imperfections of the tools (uneven tooth setting or side clearance angles), the eccentricity of the band wheels and the welding of the band. It is important to understand the exciting sources because decreasing them reduces both the vibration and deflections.

The pre-stressing force is generated with different equipment on the band saws. An old, classical solution was to use tension with a dead-load, which was theoretically perfect, but its serious drawback was that it formed its own vibration system and the pre-stressing changed periodically according to the vibration. A simple solution is an external pre-stressing screw which can set a certain pre-stress, but the system does not react to changes in tension.

Further pre-stressing solutions are the spring, the hydraulic and the pneumatic pre-stressing equipment. These respond to the changes, but they do not provide a constant pre-stressing.

The most modern pre-stressing method is the use of a servo-motor pre-stressing equipment, which keeps the pre-stressing force at a constant value with the help of a force sensor.

The pre-set tensioning force is changed during the work by the effect of the thermal strain on the band, the unloading effect of centrifugal forces and the external forces. Therefore the effective force acting on the band can be written as follows (neglecting the external forces) [1]:

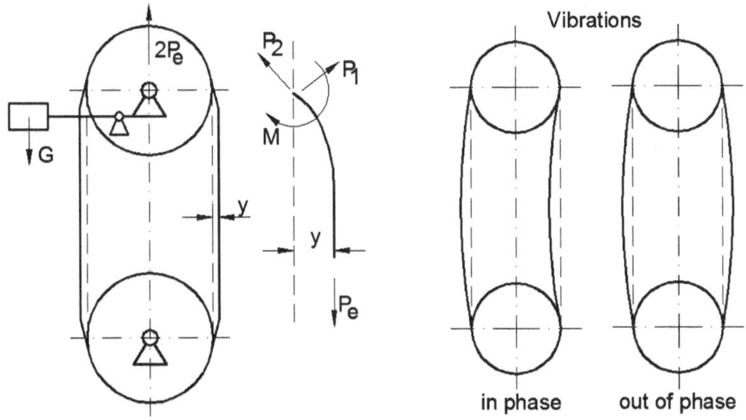

Fig. 5.2 Bending offset of a blade and vibration modes of the cutting and non-cutting spans [2]

$$N_x = \frac{P_e}{b} - \kappa\left(\rho \cdot s \cdot v^2 + \alpha \cdot s \cdot E \cdot \Delta\vartheta\right) \tag{5.1}$$

where:

P_e the pre-set tensioning force,
s the blade thickness,
v the blade velocity,
$\Delta\vartheta$ the temperature rise of the band,
α thermal expansion coefficient.

The κ strain system parameter characterises the wheel support method and the perfection of pre-stressing. For pre-stressing with dead-load is $\kappa = 0$, while $\kappa = 1$ for pre-stressing with screws. For spring, hydraulic and pneumatic pre-stressing methods the κ factor takes on intermediate values. The nearer κ is to zero, the more uniform the pre-stressing is.

The saw blade will run exactly in a plane given by the band wheels only if the bending stiffness of the blade is negligible. Otherwise the band deviates from the theoretical plane having a bending off-set [2]. The moment needed for the bending with radius R (Fig. 5.2) is:

$$M = E \cdot I \cdot \frac{1}{R}$$

which is in equilibrium with the moment $M = P_e y$

From the moment equilibrium, the band off-set is given by:

$$y = \frac{E \cdot I}{P_e \cdot R} \tag{5.2}$$

where $E \cdot I$ is the bending stiffness of the blade.

Depending on the wheel sizes and the pre-stressing, the value of the band off-set is 0.2–0.5 mm. The band off-set plays an important role in the vibration mechanism of the cutting and non-cutting spans. The greater the band off-set, the more it causes oscillating motion of the band wheels and the transfer of vibrations to the non-cutting span. This kind of vibration is referred to as coupling vibration [2]. The shape of vibrations may be in-phase or out phase, in some cases the superposition of the two modes occurs (Fig. 5.2). The in-phase blade vibration is always associated with band wheel oscillation.

This was observed experimentally. When the cutting span was disturbed, vibration displacements were also recorded in the non-cutting span.

There is a phase correlation of 90° between the vibrations in the two spans. Furthermore, the vibration frequency of both spans is identical and we see a high frequency vibration modulated by a low frequency oscillation. The modulation frequency depends on the tensioning (stretching) of the saw blade: increasing tension decreases the modulation frequency. This modulation phenomenon means a propagation of vibration energy between the two spans.

The stretching of the cutting and non-cutting spans is not exactly the same. There is a difference between the high frequencies in the cutting (ω_1) and non-cutting spans(ω_2). The average frequency is

$$\omega_{av} = \frac{(\omega_1 + \omega_2)}{2}$$

and the modulation frequency has a value of

$$\omega_{\text{mod}} = \frac{(\omega_1 - \omega_2)}{2}$$

and further

$$\omega_1 = \omega_{av} + \omega_{\text{mod}} \quad \text{and} \quad \omega_2 = \omega_{av} - \omega_{\text{mod}}$$

The high frequency oscillations on the two spans have a 90° phase shift, i.e. one vibrates with $\cos(\omega_{av} \cdot t)$, while the other with $\sin(\omega_{av} \cdot t)$. The in-phase vibration of the blade is associated with wheel oscillations, while the out-of-phase vibration is not related to wheel oscillation.

The natural frequency of the band saw blade depends on its stretching and on stresses induced in the blade by different roll tensioning methods. Tilting of the upper wheel causes a plane bending moment in the blade which adds additional tension stresses in the region of cutting edge. Wheel crown and blade overhang on the wheel can cause similar bending moments. These axial stress components are demonstrated in Fig. 6.14 (see later).

The stress due to roll tensioning generally has values of 5,000–7,000 N/cm^2 (in Sect. 6.2). The additional stress due to tilting of the upper wheel can be approximately calculated in the following way. As a consequence of tilting, the upper edge of the wheel shifts upwards with a value ΔR and its relative value related to the wheel radius R can be expressed as

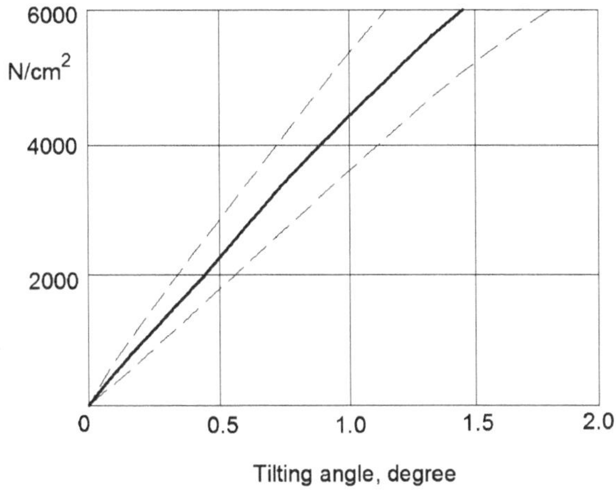

Fig. 5.3 Tilting stresses in a band saw

$$\frac{R + \Delta R}{R} = \cos\beta + \frac{\left(e + \frac{b}{2}\right)}{R} \cdot \tan\beta$$

where β is the tilting angle, b is the wheel width and e is the tilting eccentricity related to the centre line of the wheel; (the value of e may be zero).

Typical tilting stresses are given in Fig. 5.3 as a function of the tilting angle. The stresses increase almost linearly with the tilting angle. The combined stresses at the gullet bottom line are around 15,000 N/cm^2.

The natural frequency of a saw blade will be influenced by the following parameters:

The speed of the band. It only slightly influences the natural frequency. With the increase of speed, the natural frequency decreases 5–7 %. in the 0–50 m/s speed range.

The pre-stressing of the band. Pre-stressing has a determinant effect both on the transverse, and torsional vibration frequency. The transverse vibration frequency of a log rip band saw ($b = 27.5$ cm, $s = 1.5$ mm, $L_e = 60$ cm) for different pre-stressing is shown in Fig. 5.4 [1]. The Figure also shows the the effect of the peripheral speed.

The method of the pre-stressing. The method of pre-stressing may compensate the unloading of centrifugal force and temperature in different ways. An insufficient compensation decreases the natural frequency, and increases the vibration amplitudes.

The effect of the κ factor on the transverse and torsional natural frequency is shown in Fig. 5.5 depending on the peripheral speed [1].

Because a constant force wheel support system (servomotor) is expensive, it is used only on big machines. The pneumatic pre-stressing is used on small, but high quality machines, for which $\kappa \cong 0.2$. A spring-supported wheel system has values for κ around 0.5.

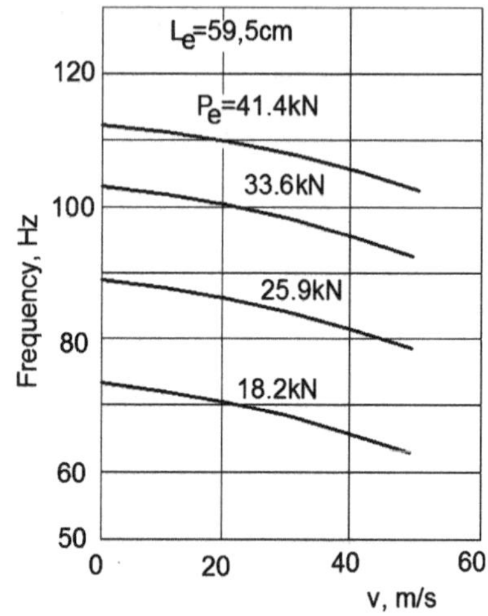

Fig. 5.4 The effect of pre-stressing of the band on its natural frequency [1]

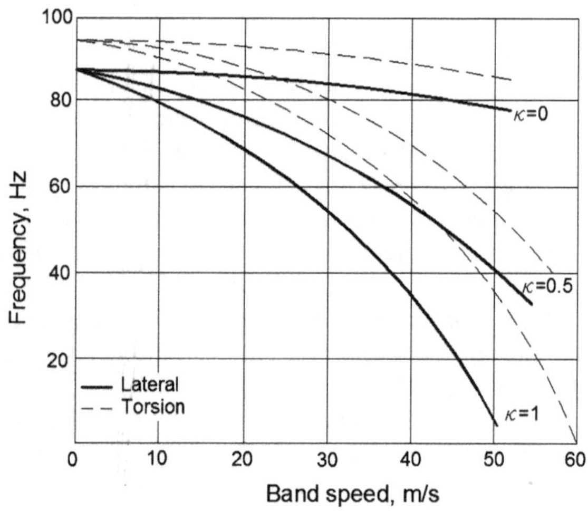

Fig. 5.5 The effect of pre-stressing methods on the change in natural frequencies [1]

The free span length of the band. An increasing free span length always decreases the natural frequency. The change of the natural frequencies for two free span lengths depending on the pre-stressing is shown in Fig. 5.6 [1].

An increase of the thickness and width of the saw blade decreases the natural frequency. Therefore it is wise to use thin blades. In the thin blade the bending stresses are smaller $\left(\sigma_m = \frac{E \cdot s}{2R}\right)$, and therefore greater pre-stressing force can be used to increase the natural frequency.

Artificial pre-stressing. It can be induced in the blade with roll tensioning so that tensile stress will remain on the edge part and compressive stress on the midsection. Straining the band causes parabolic stress-distribution so that we get the greatest resultant tensile stress at the edges. This combined straining hardly

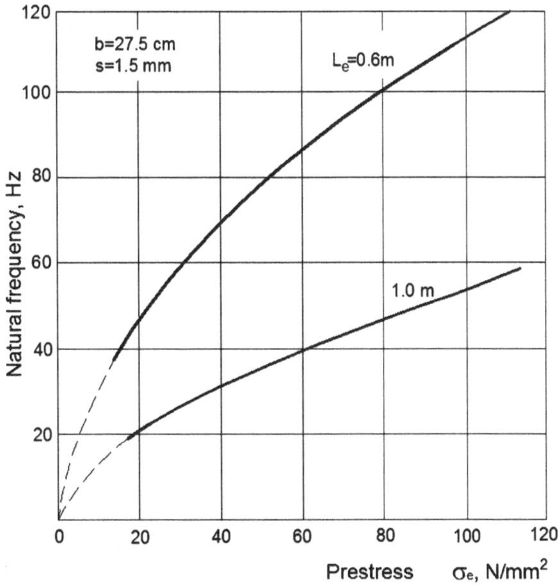

Fig. 5.6 The effect of the free span length of the band on the natural frequency [1]

influences the transverse vibrations, but it increases the torsional natural frequency by 10–20 %.

The blade heats up during the work due to friction, and a parabolic temperature distribution develops along the blade width. The exponent of the parabola changes between 2 and 3. The temperature increase causes the band to stretch in the tooth-zone, and a local decrease of the pre-stressing. This considerably lowers the torsional natural frequency.

The change in stress due to temperature variations is given by the following equation:

$$\Delta\sigma = -\alpha \cdot E\Delta\vartheta$$

where α is linear expansion coefficient (for steel $\alpha = 12 \cdot 10^{-6}$) and $\Delta\vartheta$ is the temperature increment. This thermal stress always decreases the local tension stress induced in the saw blade. For example, if $\Delta\vartheta = 60\,°\mathrm{C}$, then the thermal stress is around 15,000 N/cm^2, the same magnitude as the induced total tension stress in the gullet bottom line! Therefore, the thermal stresses considerably influence the natural frequency and especially the stability of saw blades.

The unloading of the edges of the blade primarily causes torsional deflections, decreasing the natural torsion frequency. The decrease depends on the temperature difference between the two edges of the blade. At 100 °C difference the torsional natural frequency decreases by 12–15 %, while the transverse natural frequency hardly changes, provided that the average stress of the blade remains constant.

Furthermore, it is observed that the spectra of the blade deflection show a $1/f$ behaviour, where f is the frequency. That means that the deflection amplitudes are inversely proportional to the frequency of their occurrence.

With the tilting of the upper leading wheel, asymmetric stress distribution can be generated in the blade. The direction of the tilting causes greater than the average stresses in the tooth-zone and smaller stresses on the back edge.

The extent of the tilting can be characterized with the $(\sigma_{max}/\sigma_{min} - 1)$ value and the change of the natural frequency also depends on this. The tilting slightly increases the natural torsional frequency and decreases the natural transverse frequency in a similar extent. Therefore, the tilting of the leading wheel does not significantly reduce vibration, but improves the stability of the blade on the wheel.

5.2 The Vibrations of Circular Saws

The circular saw is one of the most frequently used woodworking machines. Owing to its simple and cheap construction and relatively great sawing capacity, its use is spreading again in log sawing. (Overseas in the 19th century, logs were cut with circular saws).

An objection against circular saws was their high width of cut, causing higher kerf loss than other saws. This objection is no longer valid today. Reducing the amplitudes of vibrations, and improving the blade stability, thin circular saw blades can be produced now (e.g. 0.8 mm blade thickness and 1.3 mm kerf width up to 400 mm diameter) [3].

Thin circular saw blades require the avoidance of the critical speed, the proper clamping or guiding of the blade, running the blade in its exact plane, the accurate grinding of the teeth and roll tensioning of the blade.

Saw blade vibrations increase the actual kerf loss, reduce product accuracy and diminish surface quality. Therefore, efforts should be made to reduce saw blade vibrations to an appropriate level.

Today thin circular saw blades are generally held by a clamping plate. The larger the diameter of the clamping, the more stable the circular saw is, but this prevents cutting thicker logs. In the last decades different blade guiding systems have been successfully used, where two or more pairs of frictional pads guide the blade. The blade can move on its axis (Fig. 5.7), so it is self-adjusting.

The forms of vibration which develop in the rotating disk can be seen in Fig. 5.8 [3]. The different vibration forms are characterized by the number of nodal circles and nodal diameters. With a zero nodal circle and zero nodal diameter, the total periphery of the disk bends to the right or left. With a nodal diameter of one, half of the periphery bends, and with a nodal diameter of two, a quarter of the periphery is deflected to the right or left. In larger diameter blades, a nodal circle can develop, when another wave emerges in the inner part in the opposite phase.

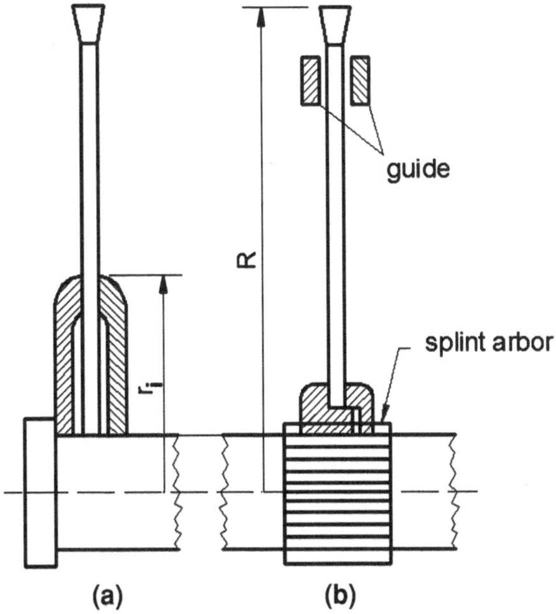

Fig. 5.7 The clamping (**a**) and guiding (**b**) of circular saws

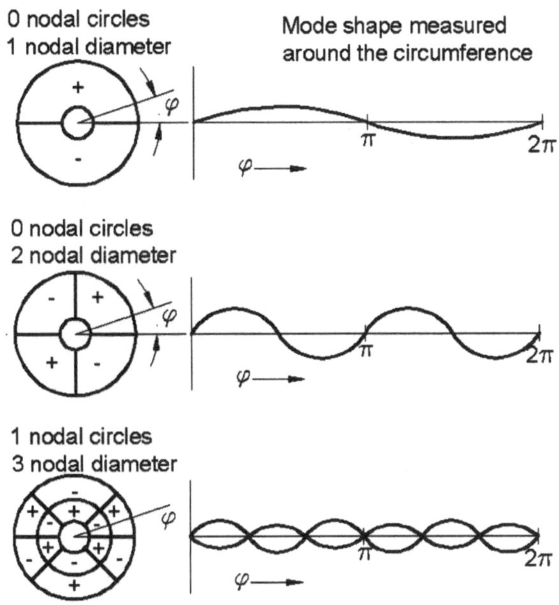

Fig. 5.8 Schematic of vibration modes of a rotating circular saw [3]

The vibration waves propagate along the periphery of the disk to the right and left. In a rotating disk, the travelling speed of the waves in the direction of rotation increases with the rotation speed (forward travelling wave), while the backward travelling wave speed decreases. If the frequency of the rotation parallels the frequency of the vibration, then a standing wave occurs on the disk. Here the rotation speed is called the *critical speed* [4].

When the operating speed approaches the critical speed, the blade becomes unstable. The large blade deflections make it impossible to produce quality sawing. The recommended maximum operating speed is at least 15 % below the critical speed [5, 6].

The frequency of the basic vibration mode can be calculated from the following expression:

$$f = \frac{\omega_0}{2\pi} = \frac{K^2 \cdot s}{4R^2 \cdot \pi}\sqrt{\frac{E}{3\rho(1-\upsilon^2)}} \tag{5.3}$$

or using the characteristic values of steel for E, ρ and v, we obtain

$$f = 253 \cdot \frac{K^2 s}{R^2} \tag{5.3a}$$

where the blade thickness s and the radius R must be substituted in Meters.

The constant K is a function of clamping ratio r_i/R in the following form

$$K = 1 + 5 \cdot \frac{r_i}{R}$$

For nodal diameters higher than 1, a similar equation can be used, but the constant K in Eq. (5.3a) will be changed. For clamping ratios between 0.25 and 0.3 the following constant can be used [3, 7]:

Nodal diameter	1	2	3	4
Constant K	253	422	717	1,223

Centrifugal forces occur in a rotating disk and these forces generate tensile stresses in both radial and tangential directions. Therefore, the natural frequency of a rotating disk slightly increases as a function of rotation speed (shown in Fig. 5.9) and it can be calculated in the following manner:

$$f_\omega = \sqrt{f_n + \lambda_n \omega^2}$$

where f_n is the natural frequency at rest for a given nodal diameter n and λ_n is a constant for the nodal diameter in question. Values for λ_n are given below for clamping ratios of 0.25–0.3 [7].

Nodal diameter	1	2	3	4
Constant, λ_n	1.25	2.3	3.75	5.5

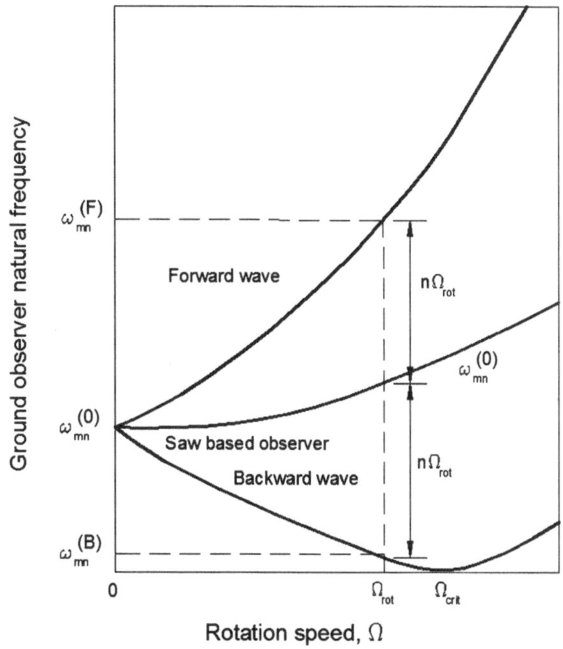

Fig. 5.9 Frequency-rotational speed diagram showing forward and backward travelling waves and the condition of standing wave resonance or critical speed [3]

The above frequencies are the "true natural frequencies," which could be seen by an observer at the saw.

The vibration waves propagate in opposite angular directions around the saw blade. Therefore, a ground based observer (or at the position of the workpiece) sees forward and backward travelling waves corresponding to a given mode n:

$$\omega_{n1} = \omega_n + n \cdot \Omega \quad \text{forward wave}$$

$$\omega_{n2} = \omega_n - n \cdot \Omega \quad \text{backward wave}$$

where Ω means the angular velocity of the saw blade, ω_n is the angular speed corresponding to the natural frequency.

The blade rotation decreases the angular speed of the backward wave and, therefore, ω_{n2} can approach the value of zero. This is the critical speed of the blade and the ground based observer sees a standing wave [3, 4]. If a standing wave occurs, the external forces always excite the blade edge at the same point which leads to an accumulation of energy causing instability (large deflections) of the blade. Therefore, the operating speed of the saw must be reduced at least 15 % below the critical speed [5, 6].

If the blade has edge slots a standing wave cannot be established at the rim of a blade and this reduces the amplitude of the unstable motion. This means, resonance and critical speed can still occur, but the slots inhibit the formation of standing waves [3].

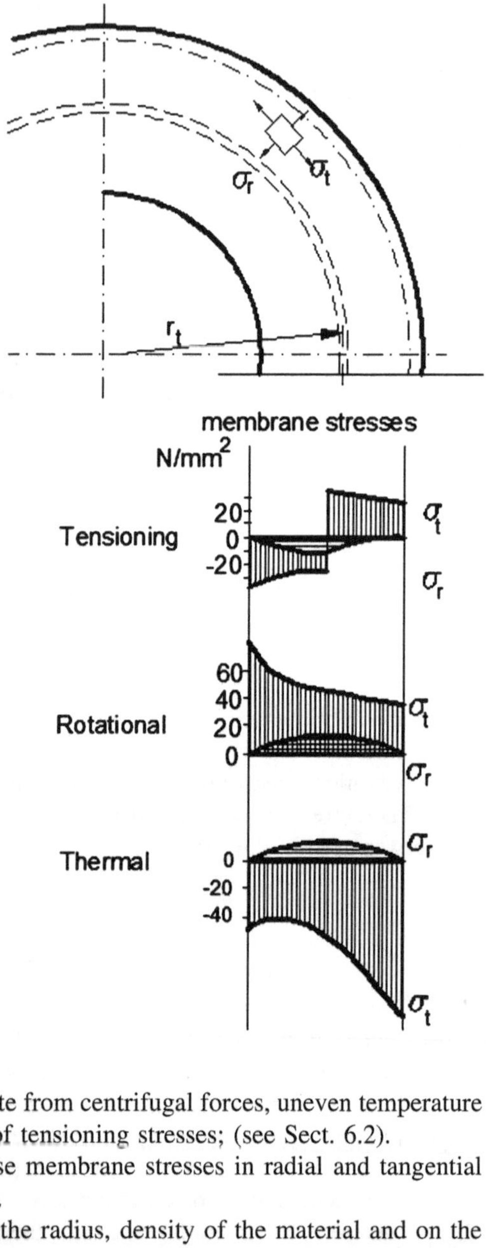

Fig. 5.10 Membrane stresses in a circular saw [3]

Stresses induced in a disc originate from centrifugal forces, uneven temperature distribution and from introduction of tensioning stresses; (see Sect. 6.2).

The common distribution of these membrane stresses in radial and tangential directions is shown in Fig. 5.10 [3].

The rotational forces depend on the radius, density of the material and on the squared angular speed. The radial and tangential stresses induced in the disc also depend on the clamping ratio. The typical stress distribution can be seen in Fig. 5.11. The radial stress increases from zero on the outer edge towards inside depending on the clamping ratio.

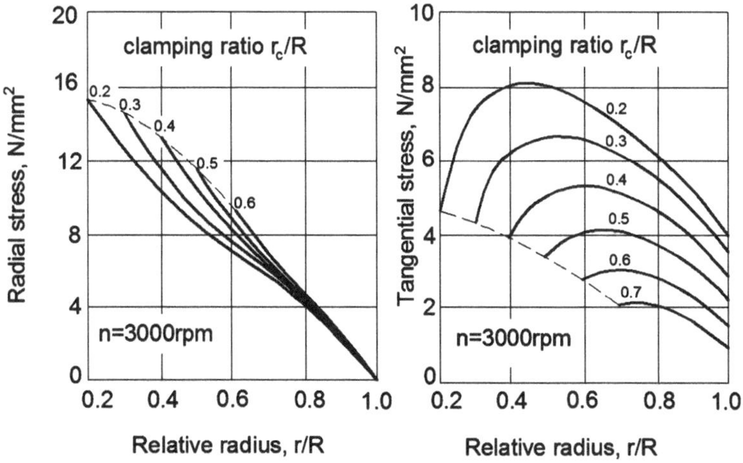

Fig. 5.11 Rotational stresses in a circular saw [8]

The tangential stress at the rim is not zero and its value increases as the clamping ratio decreases. It is very important that the rotational stresses are always tensile and contribute to the stability of the blade.

The thermal stresses depend on the coefficient of thermal expansion and the temperature distribution induced by sawing. The heat generated by friction occurs on the cutting edge and on both side of the gullet face so the disc temperature generally increases towards the outer edge in the following general form:

$$\vartheta(r) = \vartheta_0 + \Delta\vartheta\left(\frac{r}{R}\right)^n$$

where

r/R means the relative radius of the sawblade,

ϑ_0 is the ambient temperature and the exponent n varies between 2 and 3.

The radial thermal stresses increase almost linearly towards the centre of rotation, while the tangential stresses have their maximum value along the rim of the disc; (Fig. 5.12). The radial stresses are again tensile, but the large tangential stresses are compressional which may cause a corrugated edge and the disc loses its stability. Therefore, excessive thermal gradients in the disc must be avoided.

Tensioning stresses, generally induced by rolling (see Sect. 6.2), should decrease or prevent excessive tangential stresses which would cause a lower natural frequency and make the disc unstable.

The combined effect of thermal stresses and rotational stresses is demonstrated in Fig. 5.13 [7]. Temperature differences strongly lower the natural frequency for $n \geq 2$ nodal diameters. At the same time, an increasing rotational speed increases the allowable temperature difference.

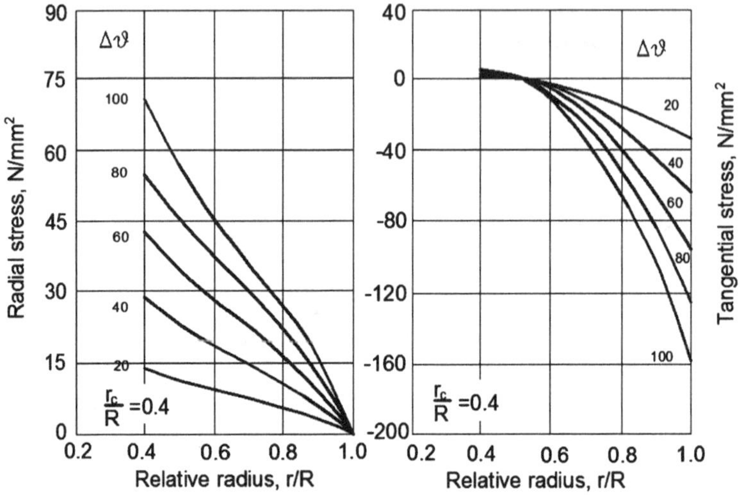

Fig. 5.12 Thermal stresses in a circular saw [8]

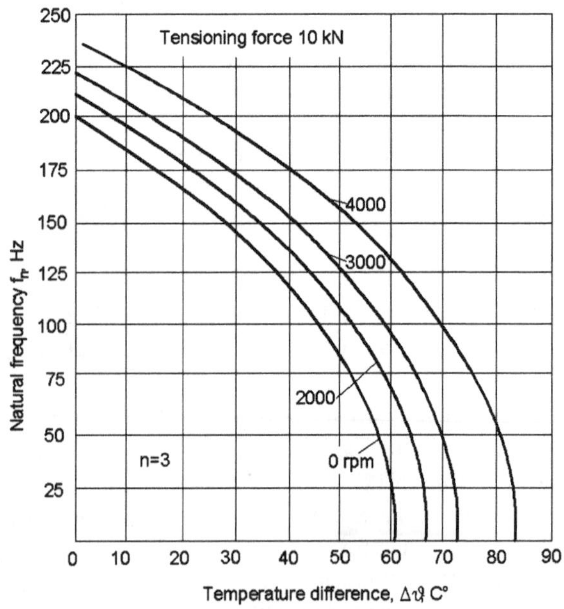

Fig. 5.13 Effect of temperature differences and rotational speed on the third natural frequency for a circular saw having $D = 400$ mm and clamping ratio of 0.3 [7]

Experimental measurements on a circular saw 500 mm in diameter and 2.5 mm thick are shown in Fig. 5.14 [6, 9]. The critical temperature difference, at which the natural frequency tends toward zero, is around 50 °C. The critical temperature difference can be estimated by the following equation [7]:

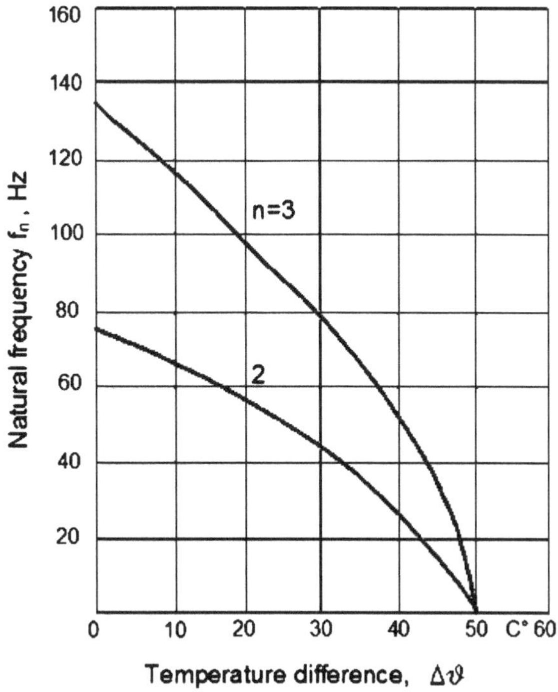

Fig. 5.14 Effect of temperature differences on the second and third natural frequencies of a circular saw with $D = 500$ mm and $s = 2.5$ mm [6, 9]

$$\Delta\vartheta_{cr} = A\frac{1}{\alpha}\left(\frac{s}{R}\right)^2$$

where α is the thermal expansion coefficient, s is the thickness and R is the radius of the saw. The constant A is a function of clamping ratio and may be substituted in the range of 0.25–0.3, as $A = 5.0$–5.5.

Thermal stresses increase the natural frequencies for the modes $n = 0$ and $n = 1$.

Roll tensioning of circular saws increases the natural frequencies for the modes $n \geq 2$ and decreases it for the modes $n = 0$ and $n = 1$ (Fig. 5.15) [7]. Generally the influence of tensioning is higher for small clamping ratios. For a given value of the rolling force, for the modes $n = 0$ and $n = 1$, the saw loses its elastic stability and buckling (dishing) occurs. That is the *critical rolling force*, at which the disc is no longer flat but forms a dish. The optimum rolling radius, for the highest possible natural frequency, is generally around 0.7 R.

Slots at the edge of circular saws are often used and they have a significant effect to decrease the lateral deflection of the blade, especially when the saw blade subjected to excess heat load from friction at the edge. The slots make it possible to eliminate the tangential stresses allowing the free expansion of the edge zone.

There are generally between 4 and 6 slots, open to the rim, and the length of a slot is related to the radius of the saw blade (ratio of slot length).

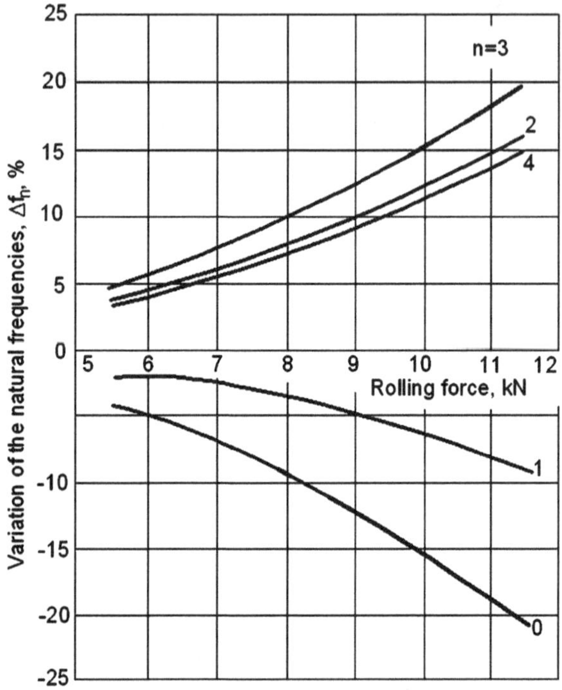

Fig. 5.15 Variation of natural frequencies as a function of tensioning force [7]

Without rim heating, slots are not beneficial. They lower resonant frequencies in modes with two or more nodal diameters. This can clearly be seen in Fig. 5.16 [10]. In the presence of rim heating, an optimum slot length ratio at about 0.3 was determined on the basis of maximum resonant frequency in the second and third nodal diameter mode.

The rim temperature of a disk varies according to sawing conditions. Therefore, it is important to know the variation of resonant frequencies with changes of temperature within the practical range. The experimental results can be seen in Fig. 5.17 [10] for the second and third nodal diameters. A short slot gives higher resonant frequency for lower rim temperatures. For higher temperatures, a longer slot gave better results up to a relative slot length of 0.33. The Figure shows the vibration frequency equal to the rotation speed (35 Hz = 2,100 rpm) by a horizontal dotted line with intersections giving resonant vibrations. These experimental results show, that the low resonant frequencies do not allow higher rotation speeds, especially with longer slots.

The most hazardous situation to be avoided occurs when the rim temperature reaches a critical value, at which the resonant frequency falls near zero. The disk lacks any stiffness against lateral forces. In practice, the slotting of circular saws has been mainly used on carbide-tipped saws operating at lower temperatures. It may be better to make the slots shorter to allow room inside them for tensioning.

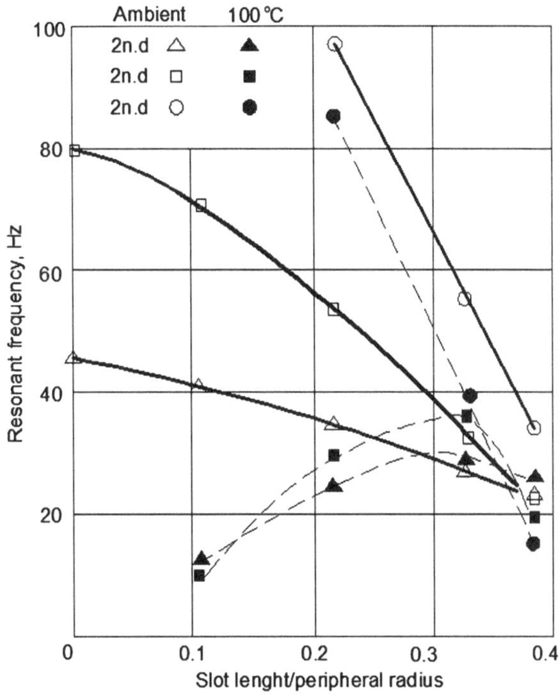

Fig. 5.16 The effects of slot length on the resonant frequencies with rim temperatures at ambient and 100 °C above ambient. Disk 460 mm in diameter, 2.4 mm thick at 2,100 rpm [10]

It was generally believed that closed slots in a disk have little effect on the operating conditions. Recently, a special closed slotting was published [11] allowing a thinner blade and a smaller kerf. The side view of the saw blade with special *S*-shaped slots is shown in Fig. 5.18 [11]. Measurements were carried out with this blade and the critical rotation speed and the critical rim temperature increased up to 60 % compared to a conventional blade.

Another interesting approach is the use of guided splined arbor circular saws (Fig. 5.19) operating at rotational speeds above the lowest critical speed [12]. This supercritical speed sawing has the potential to reduce saw kerf thickness while maintaining good sawing accuracy and feed speed. Saws with 1.5 mm kerf were used to cut 100 mm lumber at 40 m/min feed speed with sawing deviation of less than 0.25 mm.

Figure 5.20 shows the critical speed variation of a guided saw using different blade thicknesses. In practice, the saws are run between the second and third critical speeds. The speeds should be selected where the idling saws indicate the best stability.

The sawing performance is important in practice because it illustrates the relationships among feed speed, saw blade thickness and sawing accuracy. This relationship is given in Fig. 5.21 [12]. The results in the Figure clearly show the effect of both feed speed and saw plate thickness on the sawing accuracy. As could be expected, the stiffest 1.5 mm thick saw has the best sawing accuracy and can be

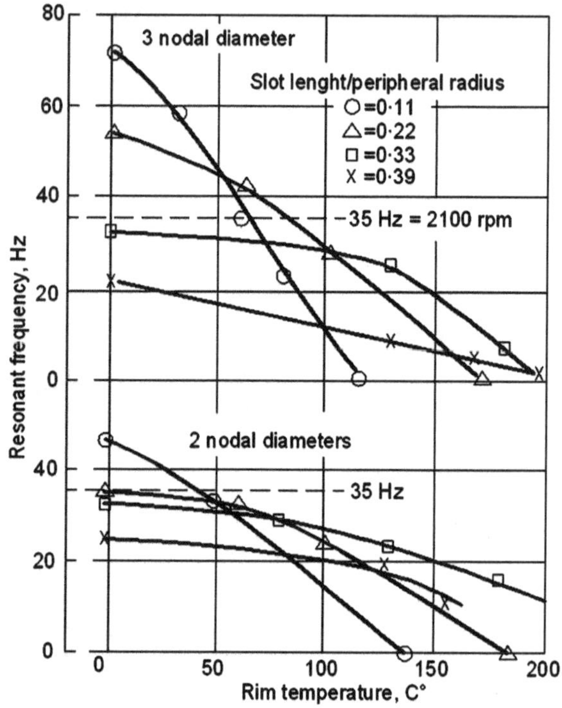

Fig. 5.17 The effect of slot length and rim temperature on resonant frequencies for a disk rotating at 2,100 rpm, five slots. Disk 460 mm in diameter, 2.4 mm thick [10]

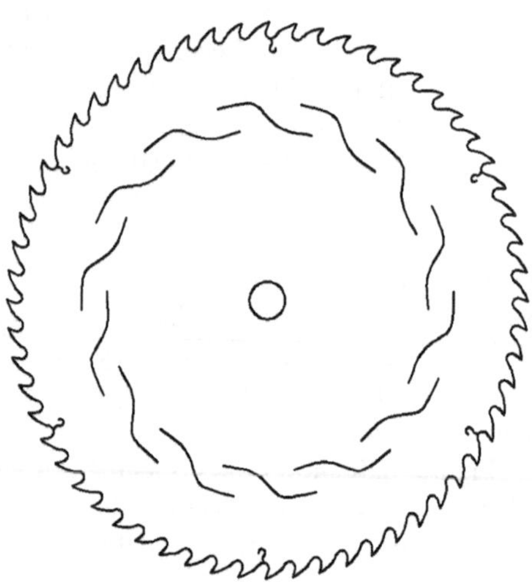

Fig. 5.18 Side view of stable saw blade with *S*-shaped slots [11]

used with a high feed speed. The 1.0 mm thin blade has higher sawing deviations and requires a lower feed speed. The maximum feed per tooth for the blade thicknesses shown in Fig. 5.21 are 0.65, 0.85 and 1.0 mm respectively.

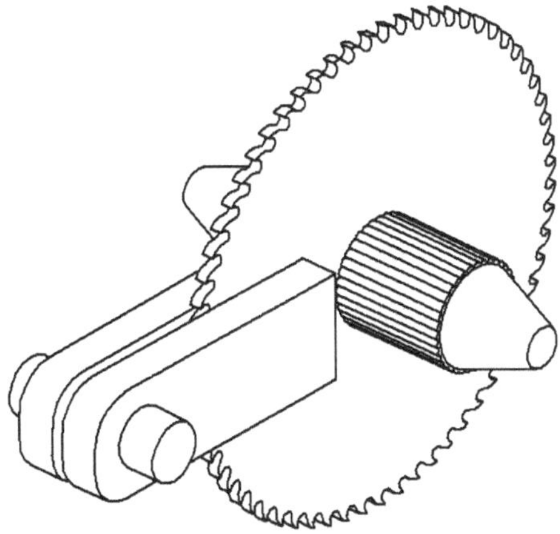

Fig. 5.19 Saw using splined arbor with guides [12]

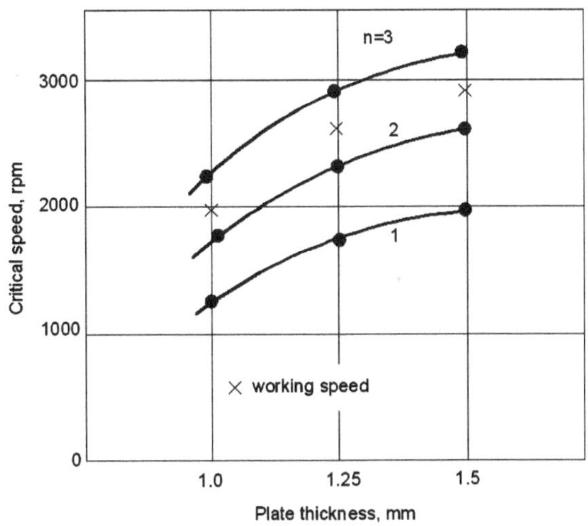

Fig. 5.20 Critical speed variation as a function of blade thickness for a guided saw 500 mm in diameter [12]

Another interesting development was the "hyper critical saw blade" with centre slots [13] to improve the stability of a thin kerf saw blade and its resistance to thermal buckling. The idea of this slot spacing was to allow the free radial expansion of the blade from centrifugal forces of rotation (Fig. 5.22).

This saw blade is also fixed by flanges from both sides. However, tightening the flange will limit the free expansion of the blade and above a certain torque, the saw blade would work as a conventional saw blade. Therefore, the flange is tightened with only enough torque to keep the saw blade stiff. The effect of free expansion on the critical rotational speed can be seen in Fig. 5.23 for a saw blade 305 mm in diameter, a 1.5 mm kerf and 0.9 mm thick blade. The Figure shows that with free expansion, the critical rotational speed is at least 50 % higher compared to a rigidly fixed flange. This saw blade was used to cut wood 60 mm thick at a rotation speed of 4,000 rpm, a feed speed of 80 m/min and a 0.5 mm feed per tooth.

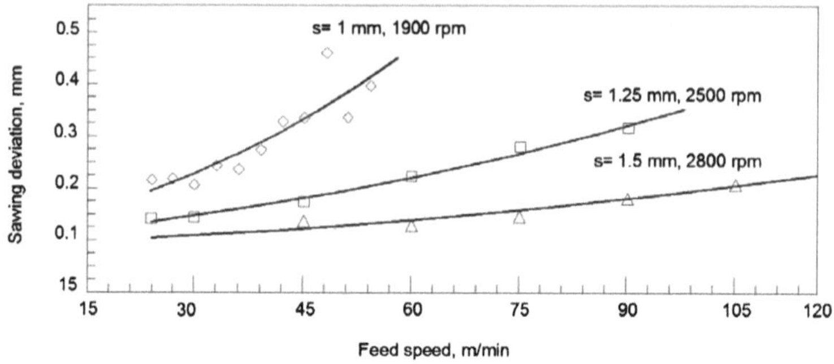

Fig. 5.21 Sawing deviation vs. feed speed for a guided saw using different blade thicknesses. Cutting height is 100 mm, with 42 teeth [12]

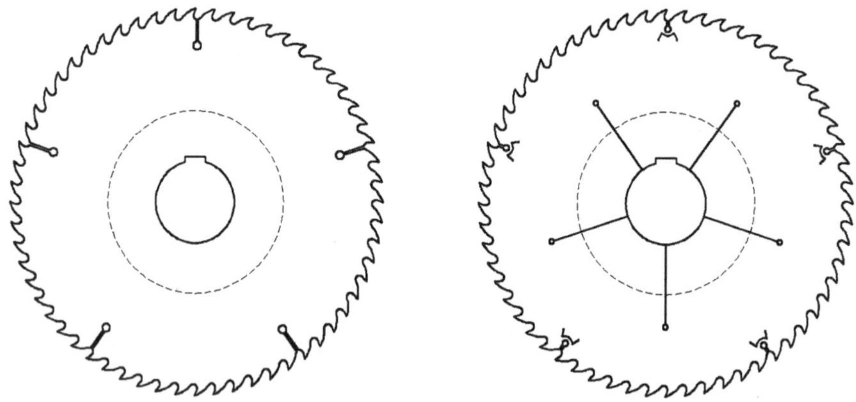

Fig. 5.22 Conventional (*left*) and hyper critical saw blade (*right*) [13]

Fig. 5.23 Relationship between tightening torque and critical speed [13]

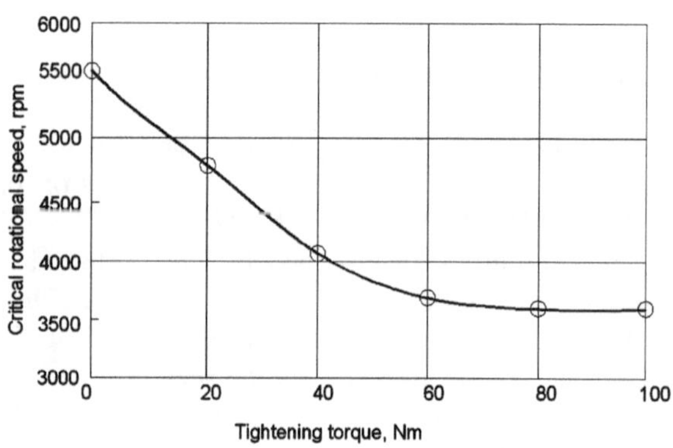

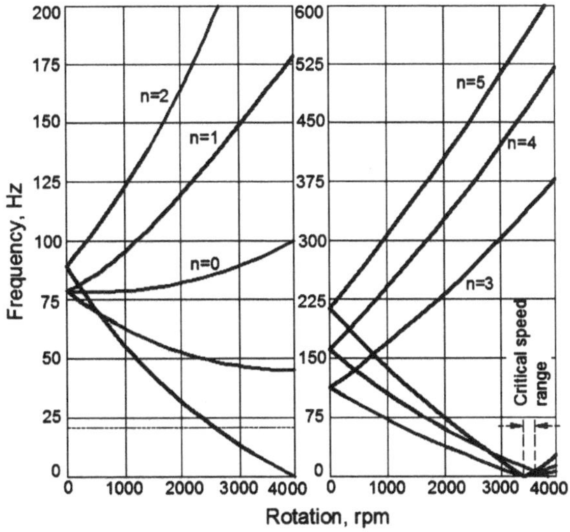

Fig. 5.24 Frequency-rotational speed diagrams showing the critical speed range [3]

In the development of circular saws, all natural frequencies are calculated and experimentally confirmed. These results give the range of critical rotational speeds which must be avoided on both sides, at least, to a distance of 15 % from the critical values. The calculations and experimental results are shown in Fig. 5.24 [3].

5.3 Washboarding

Band saws and circular saws often cause a wave like surface pattern on surface of sawn lumber which is widely known as washboarding. This phenomenon occurs when using higher blade speeds, faster feed speeds and thinner band and circular saws. A washboarding profile can be up to 0.6–0.8 mm deep on the face of the lumber. Therefore, the rough size must be large enough to remove sawing deviations and the depth of washboard. This considerably decreases lumber recovery.

Washboarding is a dynamic phenomenon occurring at high frequencies. It was observed that the main cause of washboarding is the self-excited vibration during sawing [14–16]. Furthermore, an important condition causing washboarding is when the frequency of the tooth passage is slightly higher than the natural frequency. It is also concluded that all the washboards for band saws were produced by torsional vibration.

The typical wave like pattern of a washboard is shown in Fig. 5.25 [14].

The wave-like pattern is characterized by the wave length in feed direction (λ_x) and in the direction of cut (λ_y). The lateral tooth displacement is given by the harmonic equation

$$y = y_0 \sin(2\pi \cdot f_n t)$$

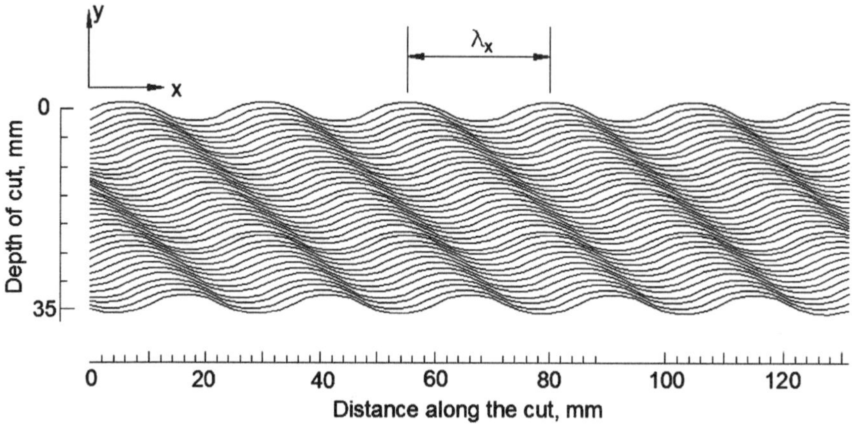

Fig. 5.25 Typical wave-like pattern of a washboarding band saw [14]

where y_0 is the maximum lateral displacement of a tooth due to self-excited torsional vibration with natural frequency f_n.

The wave length of this vibration is

$$\lambda = \lambda_y = \frac{v}{f_n}$$

where v is the band speed. The tooth frequency f_t for a given point is calculated as

$$f_t = \frac{v}{t}$$

where t means the tooth pitch. If $f_t = f_n$, then each tooth approaches a given point in the same phase. Therefore, peaks and valleys run parallel to the feed direction.

Experiments have shown that a fully developed washboarding always occurs at tooth passing frequencies that are slightly higher than the natural frequency of the blade (Fig. 5.26) [14].

The band natural frequency was 196 Hz and washboarding occurred between tooth frequencies of 196 and 204 Hz. This is a resonance phenomenon and energy will be accumulated in the subsequent cycles, which increases the lateral displacement. It may be assumed that the slight difference in the frequencies will increase the excitation forces. When there is a slight difference between the two frequencies, a small ΔT time difference occurs between the period of tooth passage and the period of band saw vibration (Fig. 5.27).

This time difference is given by

$$\Delta T = \frac{1}{f_t} - \frac{1}{f_n} = \frac{-(f_t - f_n)}{f_t \cdot f_n}$$

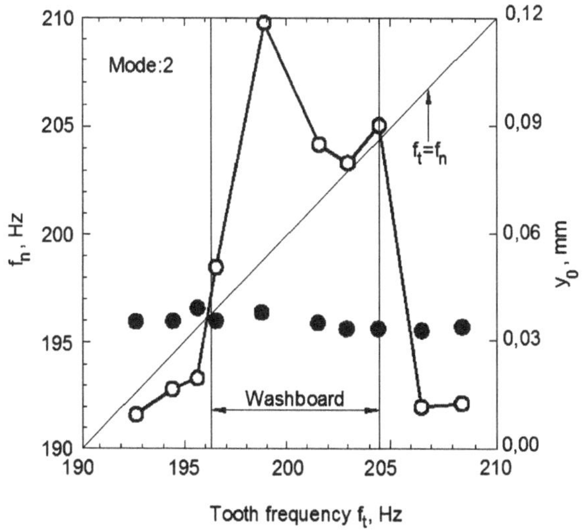

Fig. 5.26 Washboard range just after a resonance; (*black points* show resonance frequency) [14]

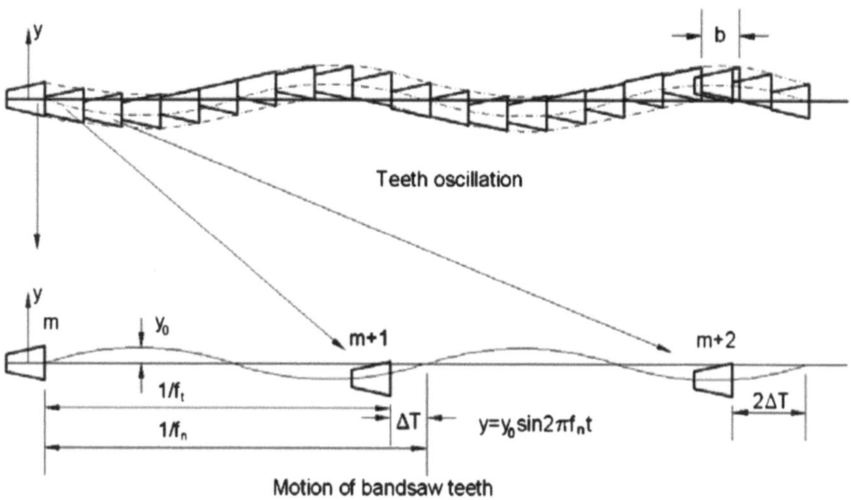

Fig. 5.27 Time difference between tooth passing frequency and blade resonant frequency [14]

and the corresponding displacement

$$\Delta y = v \cdot \Delta T = \frac{-(f_t - f_n) \cdot t}{f_n}$$

The geometry of a washboarded surface is shown in Fig. 5.28 [14].

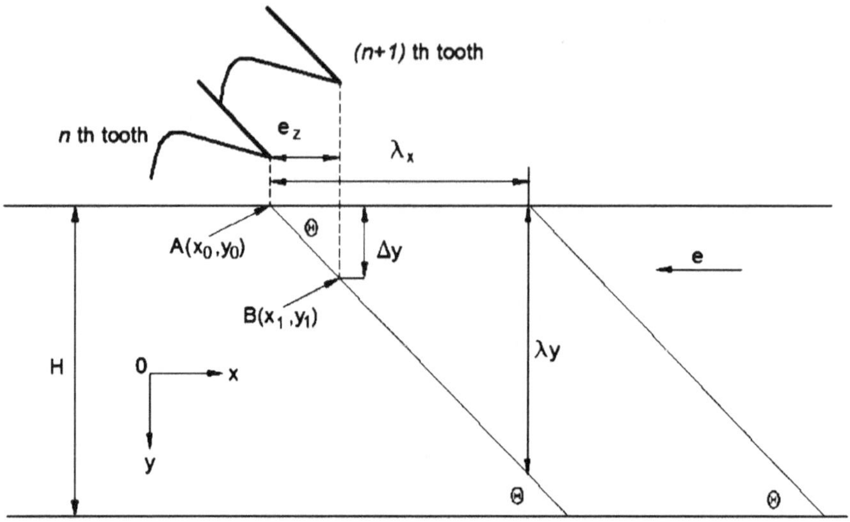

Fig. 5.28 Geometry of a washboarded surface of a workpiece [14]

Due to the frequency differences, the peaks and valleys are at angle Θ. This angle is calculated as

$$\tan\Theta = \frac{\Delta y}{e_z} = \frac{-v(f_t - f_n)}{e \cdot f_n} = \frac{\lambda_y}{\lambda_x}$$

where $e_z = e/f_t$ is the tooth feed. From the above equation

$$\lambda_x = \frac{e}{f_t - f_n}$$

If $f_t = f_n$, then $\lambda_x = \infty$ and that means that the peaks and valleys on the surface run parallel to the feed direction.

To minimize the effect of washboarding the following suggestions can be done:

Because washboarding usually occurs in a specific rotation speed range, washboarding may be eliminated by a 5–10 % increase or decrease of the blade speed. Somewhat similar effect can be achieved by increasing or decreasing of the band saw tension, which alters the natural frequency of the saw blade. A thinner blade has a lower natural frequency and is more likely to cause washboarding. Thus a thicker blade may reduce or eliminate washboarding.

Excitation forces depend on the tooth bite which could be reduced by reducing the tooth pitch.

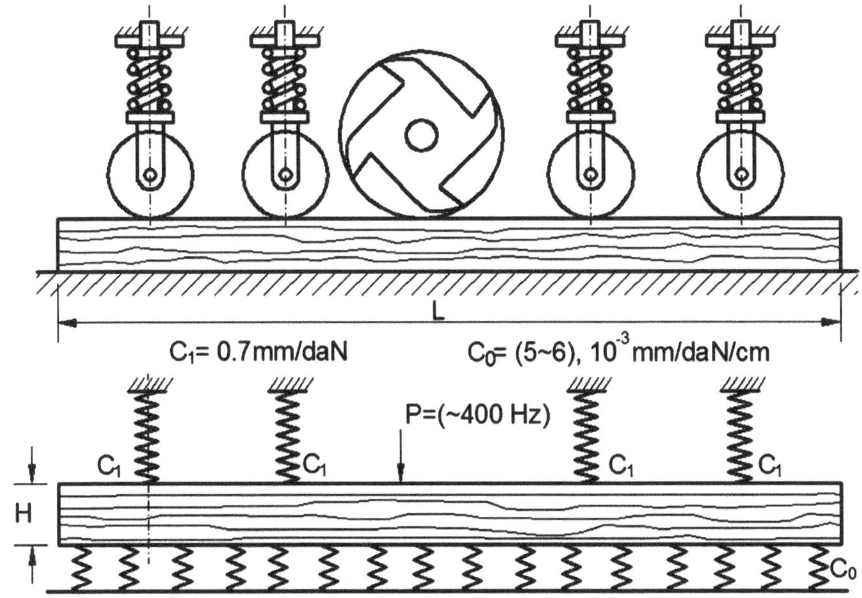

Fig. 5.29 Sketch of a multi-head moulder and its mechanical model [17]

5.4 The Vibrations of the Workpiece

The machined workpiece forms a vibration system and causes vibration displacements while running through the machine. The vibration displacements worsen the accuracy of the processing and the surface quality, so vibration deflections have to be kept as low as possible.

The vibrations of the workpiece depend on many factors. The most important are:

- the thickness and length of the workpiece,
- the exciting forces and the instantaneous boundary conditions in relation to the location of machining.

Boundary conditions mean the position of the pressing rollers, their pre-stressing and the spring stiffness.

These experimental results are from a multi-head moulder, but the relationships refer implicitly to thickness planer and shaper moulders and milling machines [17].

The principle of a multi-head moulder and its mechanical model can be seen in Fig. 5.29. The machined material is pressed to the machinetable by spring loaded rollers. The vibration mass is represented by the machined material, the spring constants on the two sides of the mass differ from each other by at least two orders of magnitude. The machinetable can be considered rigid in relation to the workpiece. Therefore, the spring constant on the side of the machine table is determined

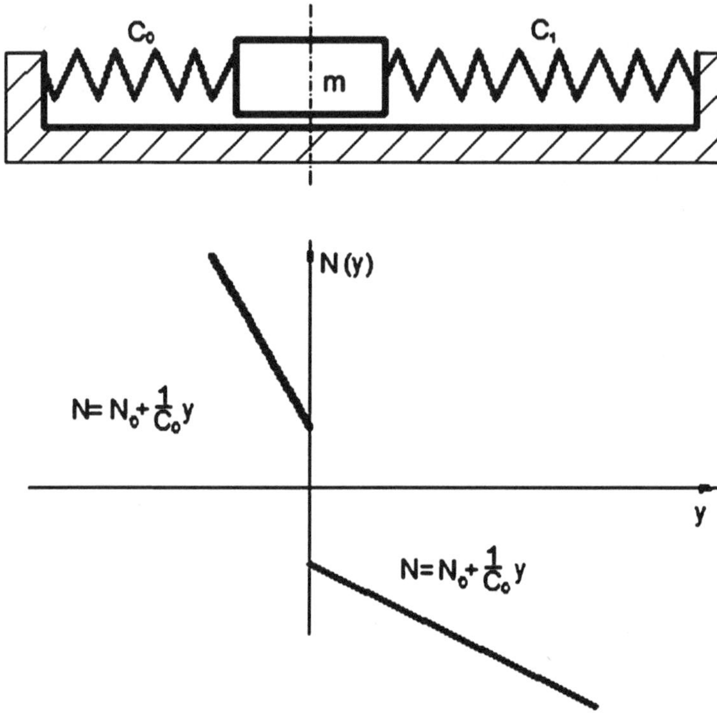

Fig. 5.30 The forces acting on the vibrating mass depending on its displacement [17]

by the elasticity of the wood. This spring constant primarily depends on the thickness of the workpiece related to the elastic modulus (*H*/*E*). On the upper side of the wood the compressive force is determined by the pre-stressing of the rollers and their spring characteristics. The frequency of the exciting forces is usually between 400 and 800 Hz. They depend on the tool rotation speed and the number of the knives, the mechanical system of the vibrating mass and the change of the forces depending on their displacement are shown in Fig. 5.30. The forces on the machine table increase very sharply. The forces on the side of the rollers vary moderately. We conclude that the vibration deflections will be highly asymmetric.

The possible vibration modes of a workpiece are shown in Fig. 5.31. They can have transverse, rotating (torsional) and bending vibrations. The first two vibration forms are characteristic of shorter and thicker workpieces. Bending vibrations are typical for the longer and thinner workpieces. Rotating vibration can appear at the entry and exit of the workpiece into and out of the machine, since at this time the force pressing down the workpiece is asymmetric.

The nodes of the bending vibration waves only are under the pressing rollers, if the spring constant of the pressing rollers is big enough providing a rigid clamping. The stiffness of the pressing rollers can be characterized with the following, dimensionless number:

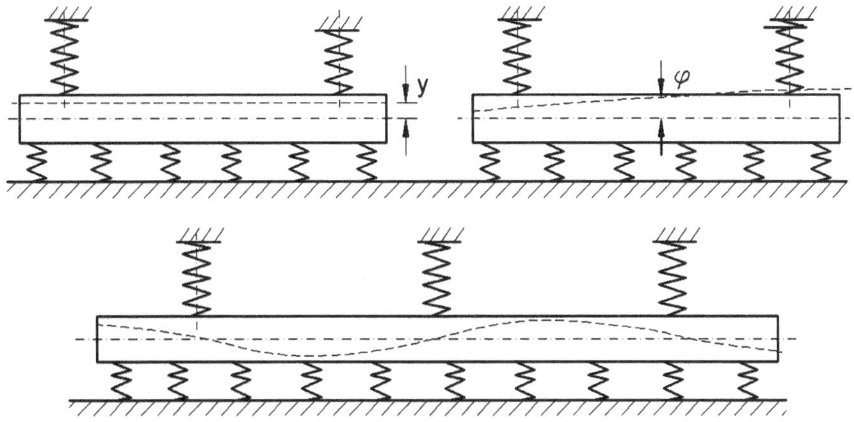

Fig. 5.31 The possible vibration modes of a workpiece [17]

$$K = \frac{c_1 \cdot E \cdot I}{l^3}$$

where:

c_1 stiffness of the spring loaded rollers,
EI bending stiffness of the machined board,
l the distance of the pressing rollers.

For multi-head moulders, K values change between 4 and 5, and this means that the frequency of the vibration depends on the length of the workpiece, not on the distance between the pressing rollers. Pressing by the rollers does not mean rigid clamping.

The exciting force is given primarily by the radial component of the cutting force. The radial component depends on the sharpness of the knife, the chip thickness and the mechanical properties of the wood. The exciting forces per single rotation for different numbers of teeth are shown in Fig. 5.32. The curve of the exciting forces can be expressed with the Fourier series expression (or FFT analysis). If we assume that the exciting force agrees with a half period (z = 8–12), then we get the following infinite series:

$$P(t) = P_0\left(\frac{1}{\pi} + \frac{1}{2}\sin\omega \cdot t - \frac{2}{3\pi}\cos 2 \cdot \omega \cdot t - \frac{2}{15\pi}\cos 4 \cdot \omega \cdot t - \cdots\right) \quad (5.4)$$

The angular frequency of the exciting forces: $\omega = \frac{z \cdot n}{9.55}$ where:

z the number of the knives,
n the rotation speed,

when n = 6,000 rpm and for z = 4–8 knives we get ω = 2,500–5,000 rad/s.

The differential equation of motion of the vibrating mass can be written as follows (see: Fig. 5.30):

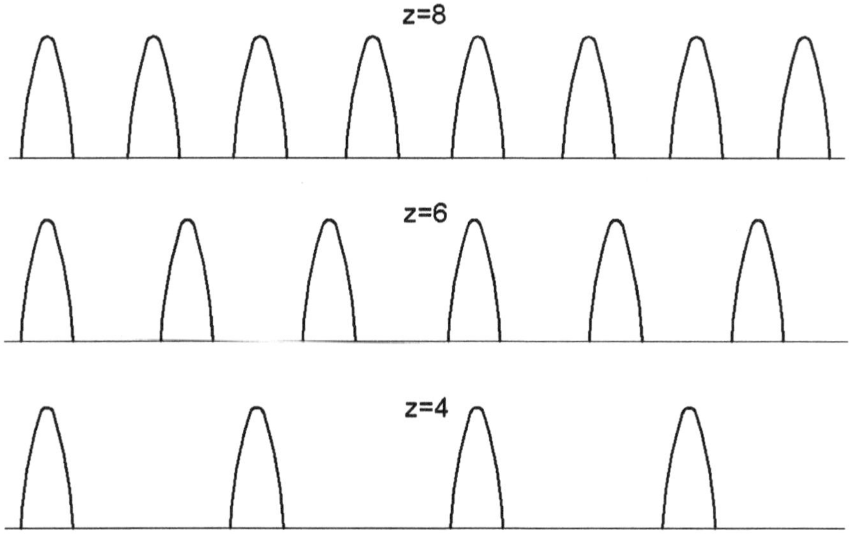

Fig. 5.32 The succeeding exciting forces with different numbers of teeth [17]

$$m\frac{d^2y}{dt^2} + \left(N_0 + \frac{y}{c}\right) = P(t) \tag{5.5}$$

The general solution of the differential equation is:

$$y = A_1 \cos\alpha \cdot t + A_2 \sin\alpha \cdot t + \frac{1}{m \cdot \alpha}\int_0^t P(t) \cdot \sin(t - \tau)d\tau \tag{5.6}$$

and $\alpha = \frac{1}{\sqrt{m \cdot c}}$

where:

A_1, A_2 integration constants,

α the angular frequency of the free vibration.

The first two parts of Eq. (5.6) describe the free vibration of the system, while the third part gives the forced vibration. The vibration is very asymmetric because the spring constants are very different on the two sides of the mass and they are not coupled to each other. We can see the developing vibration picture in Fig. 5.33, which is strongly asymmetric, and the forced vibration is superimposed on the free vibration with a substantially larger frequency. The following relation exists between the two amplitudes:

$$Y = \frac{\alpha_0}{\alpha_1}(y_0 - N_0 \cdot c_0)$$

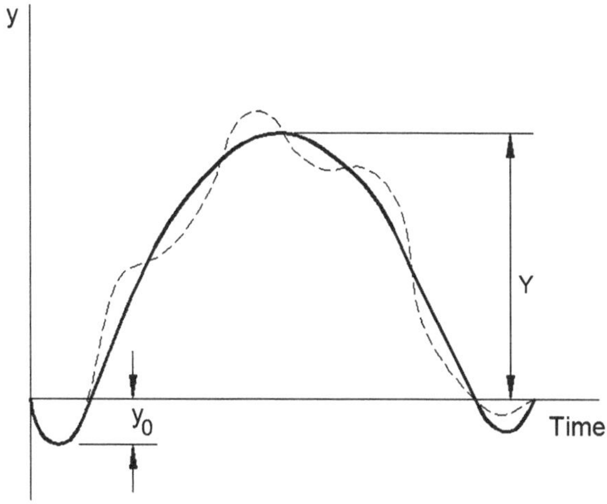

Fig. 5.33 The asymmetric vibration of the workpiece [17]

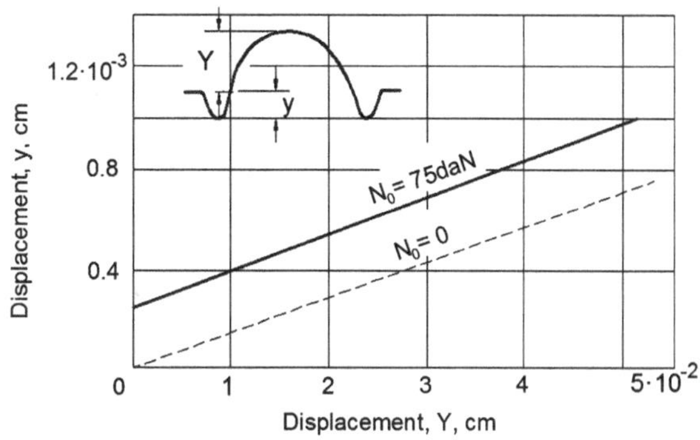

Fig. 5.34 The effect of the pre-stressing on vibration deflection [17]

where:

α_0 angular frequency of the workpiece against the machinetable,
α_1 angular frequency of the workpiece against the spring loaded rollers.

It can be seen from the equation that the Y amplitude can be decreased substantially by pre-stressing the supporting rollers (Fig. 5.34).

The amplitude of the forced vibration results from the integration of the third part of Eq. (5.6) If the $P(t)$ disturbing function is given with Eq. (5.4), then the displacement given by:

$$y = \frac{P_0}{m \cdot \alpha^2}\left[\frac{1}{\pi} + \frac{1}{1-\left(\frac{\omega}{\alpha}\right)^2} \cdot \frac{1}{2}\sin\omega \cdot t - \frac{1}{1-\left(\frac{2\omega}{\alpha}\right)^2} \cdot \frac{2}{3}\cos 2\omega \cdot t - \ldots\right] \quad (5.7)$$

Fig. 5.35 Vibration amplitudes generated by thickness planer and shaper along the length of the workpiece [17]

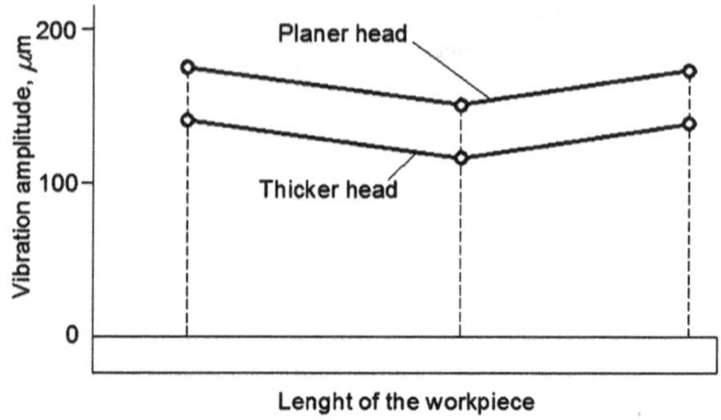

The angular frequency of the bending vibration can be calculated from the following equation:

$$\alpha_b \cong 9\sqrt{\frac{E \cdot I}{q \cdot L^4}} \tag{5.8}$$

where:

q the mass per unit length,
L the length of the rod.

The results of these calculations are summarized in the following table, which refer to a 4 × 5 × 150 cm window frame component.

The values of the vibration parameters

Parameters	Free vibration	Forced vibration
α_0	12,780 rad/s	
α_1	184.6 rad/s	
α_b	208.8 rad/s	
ω	–	3,770 rad/s
y_0	4–6 µm	
Y_{free}	110–240 µm	
Y_{for}	–	50–70 µm

Acceleration sensors were mounted on this workpiece, and the vibration amplitudes of the workpiece were measured during the processing cycle.

The vibration amplitudes are very small toward the machine table, but their frequency (α_0) is large. On the contrary, the vibration amplitudes toward the supporting rollers are much higher. It is interesting to note that the calculated transverse vibrations (α_1) and the angular frequency of the bending vibrations (α_b) do not substantially deviate from each other.

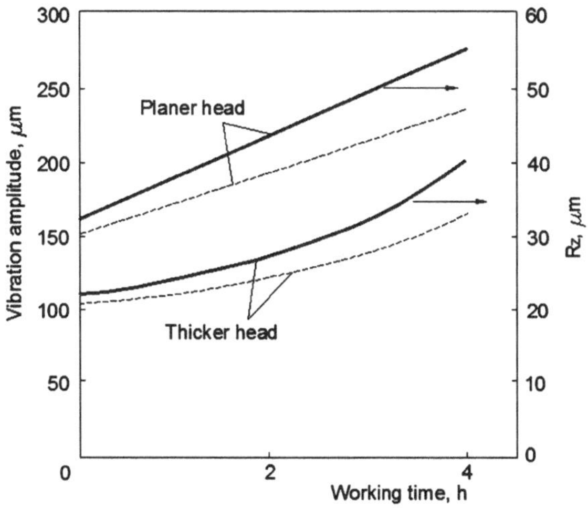

Fig. 5.36 The effect of working hours on vibration amplitudes and surface roughness [17]

The measured vibration amplitudes of the workpiece along its length are shown in Fig. 5.35. When the moulder-head is processing the front and the back end of workpiece, these sections of the workpiece fall outside the immediate effect of the pressing rollers. That increases the vibration amplitudes at the moulder-head. When the rollers support the workpiece in relation to the moulder-head centrally, then the deflections are minimized. The deflections depend on the direction of exciting forces, whether they act toward the machine table or the supporting rollers. In the latter case the deflections will be greater (e.g. in a shaper). The measured deflections are similar to the calculated values.

During the working operation, the knife becomes duller and the radial force will continuously increase. As a consequence, the exciting force increases and the deflections will also increase. This can be seen clearly for the shaper and thickness planer in Fig. 5.36.

The periodic transverse deflection of the workpiece has a substantial effect on its surface quality. We have illustrated the surface roughness together with the deflections in Fig. 5.36 (continuous lines). The surface roughness increases similar to the vibration amplitudes.

Further measurements have shown that there is a close correlation between the vibration amplitudes and the surface quality. The correlation can be seen in Fig. 5.37. A dull knife produces a slightly worse surface quality at the same amplitude, as the edge of its larger radius causes more damage to the surface. But the definitive factor is always the vibration amplitude. The surface quality is not constant along the length of longer workpieces. It is always the best in the middle parts, where the vibration deflections are the smallest.

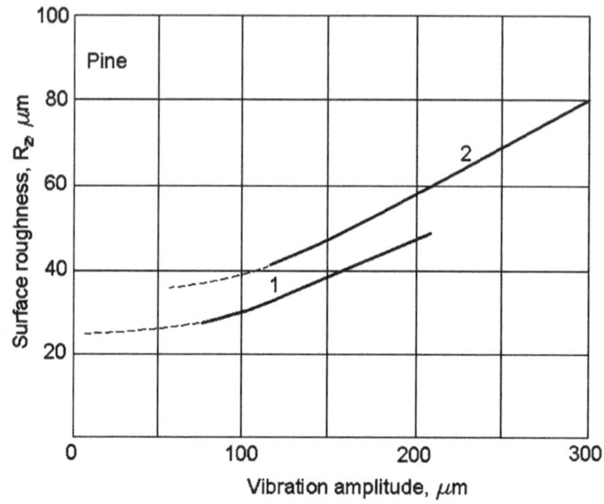

Fig. 5.37 Influence of vibration amplitude of the workpiece on the surface roughness. (*1*) sharp tool; (*2*) blunt tool

5.5 Pneumatic Clamping of a Workpiece

Computer controlled routers and machining centres represent the leading woodworking technology. In order to achieve an appropriate dimension accuracy and maximum quality, the workpiece has to be fastened on the machinetable securely. The most frequent clamping method for workpieces with a flat surface is a vacuum field. The vacuum pressure generates a vertical force downward, and the horizontal friction force prevents the workpiece from shifting in any direction. The surface of the clamping table contains grooves arranged in a checkered pattern. The vacuum area of the desired size and shape may be formed by placing rubber profiles in the grooves. The vacuum pump generally allows a maximum vacuum pressure of 0.8–0.9 bar. The pressing force generated by the vacuum is

$$F_v = p_v \cdot A$$

where p_v is the vacuum pressure and A is the effective area under pressure.

A workpiece clamped by vacuum on the machine table can be shifted by applying a horizontal force. Figure 5.38 shows the force and displacement relationships for a given workpiece [18].

The horizontal force shifting the workpiece is a friction force and, so the friction coefficient can be calculated:

$$F_h = \mu_0 \cdot F_v \quad \text{or} \quad F_h = \mu \cdot F_v$$

where μ_0 is the static friction coefficient for a resting body, and μ is the kinetic friction coefficient for a body is in motion.

In Fig. 5.38 we can see that the static friction coefficient is slightly higher than the kinetic friction coefficient. After a given sliding distance, the kinetic friction

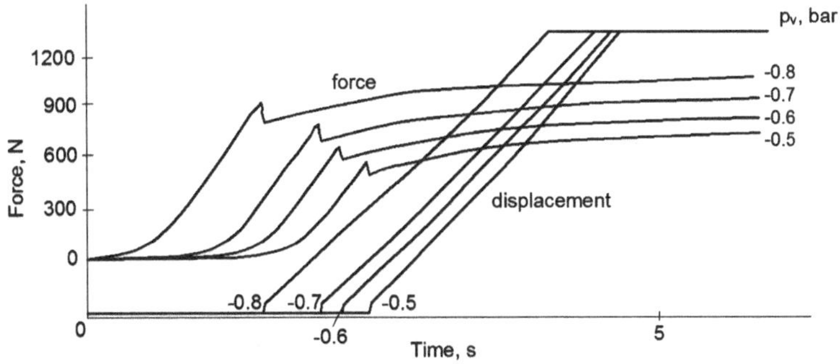

Fig. 5.38 Force requirement for shifting a workpiece on the machine table. Laminated particle board, $A = 583\ \mathrm{cm}^2$, at different vacuum values [18]

Fig. 5.39 Static friction coefficients along and across the grain for various wood materials in a vacuum range from 0.7 to 0.8 bar [18]

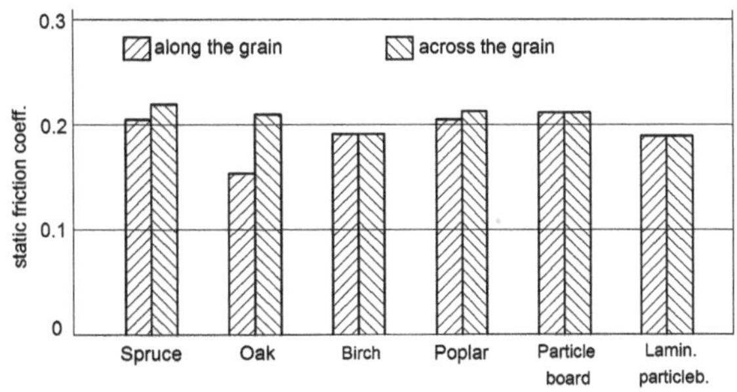

coefficient slightly increases and it can surpass the static value. From a practical point of view, the static friction coefficient is definitive.

To calculate the friction coefficient, the vertical pressing force must be known. Apparently this is easily calculated from an equation given above, but the selection of the effective surface area under suction may cause some troubles. Measurements have shown that the ratio of effective and actual surface area depends on the shape and size of the workpiece and also on the strength of the vacuum. Generally this ratio varies between 0.80 and 0.90 and the higher values are valid for larger workpieces and lower vacuum pressures [18]. For MDF, however, lower values between 0.7 and 0.8 are obtained.

The static friction coefficients for various wood species and wood products are given in Fig. 5.39 [18]. The friction coefficient along the grain is slightly less than across the grain. Laminated wood products and hard woods, such as oak, have a lower friction coefficient and are more sensitive to inaccurate clamping.

The dynamic behaviour of a clamped workpiece can be examined by impact forces made by a pendulum loading device [18]. The impact characteristics such as the deformation, force, speed and deceleration during the impact can also be calculated on an elastic material impacted by a mass. Taking a linear force deformation relationship in the form

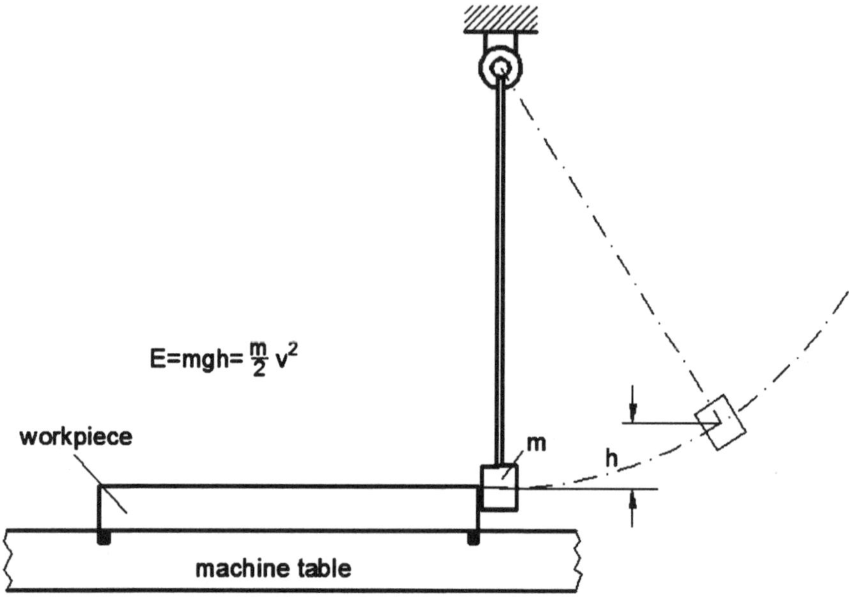

Fig. 5.40 Pendulum loading device [18]

$$F = B \cdot x$$

where B means the stiffness of the material, then the maximum deformation is given by the energy equation (see Fig. 5.40):

$$m \cdot g \cdot h = \int_0^x F \cdot dx \quad \text{and} \quad x_{max} = \sqrt{\frac{2 \cdot m \cdot g \cdot h}{B}} \tag{5.9}$$

When a mass falls onto a surface, the force is determined by the deceleration and, therefore, the following equation is valid

$$F = m \cdot a = B \cdot x$$

and

$$a = \frac{B \cdot x}{m}$$

where a is the acceleration.

After contacting the surface, the speed of the pendulum mass decreases. At the beginning of the impact the approach speed is

$$v_0 = \sqrt{2 \cdot g \cdot h}$$

where h is the falling height (see Fig. 5.40). As the deformation reaches its maximum value, the speed tends toward zero.

The change in speed can be calculated in the following way:

$$v_x dv = -a \cdot dx = -\frac{B}{m} x \cdot dx$$

after integration we have

$$\frac{v_x^2}{2} = -\frac{B}{m}\frac{x^2}{2} + C$$

The initial condition is taken as $x = \rightarrow v_x = v_0$, therefore,

$$C = \frac{v_0^2}{2}$$

and further

$$v_x = \sqrt{v_0^2 - \frac{B}{m} x^2} \tag{5.10}$$

If the deformation x approaches its maximum value x_{max}, then the speed of the mass tends toward zero.

Using the last equation and keeping in mind that $v = \frac{dx}{dt}$, the impact duration can be determined:

$$\Delta t = \int_0^{x_{\max}} \frac{dx}{v_x} = \frac{\pi}{2}\sqrt{\frac{m}{B}} = \frac{\pi}{2}\frac{x_{\max}}{v_0} \tag{5.11}$$

The maximum force is given by the following equation

$$F_{\max} = B \cdot x_{\max} = \sqrt{2 \cdot B \cdot m \cdot g \cdot h} \tag{5.12}$$

The above equations are valid if the counter-force is greater than the maximum impact force. If not the impacted body will slide to a given distance and the surplus energy of the mass will be consumed by friction. It is important to note that a considerable part of the impact energy is converted into deformation and acceleration of the body. Theoretically, a pure elastic deformation does not consume energy and the impacting mass would be returned to its initial position.

The displacement of an oak specimen during impact loading is shown in Fig. 5.41 [19]. The impacting mass was 10 × 10 × 10 cm, 7.8 kg and a drop height of 7.5 cm. The impact energy is 5.74 Nm and the impact speed is 1.21 m/s. The friction force of the specimen was appr. 700 N. Friction energy of 0.812 Nm was consumed over a sliding distance of 1.16 mm and this amount is only 14 % of the total impact energy.

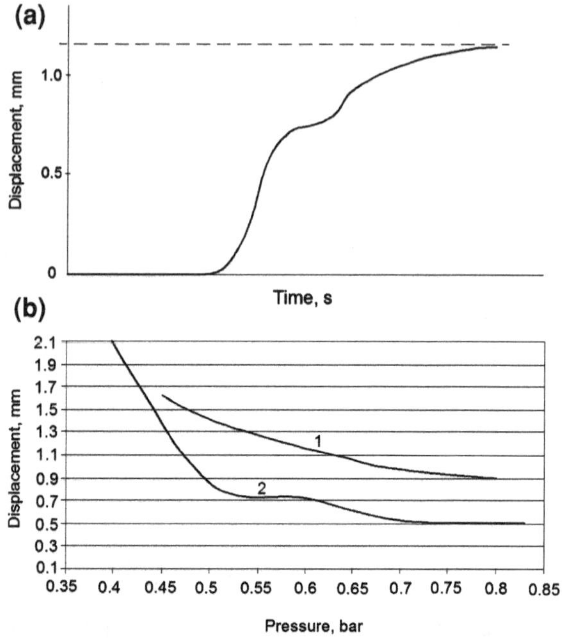

Fig. 5.41 Displacement of an oak specimen under impact load (**a**); and displacement of oak (*1*) and chipboard (*2*) specimens as a function of clamping pressure (**b**)

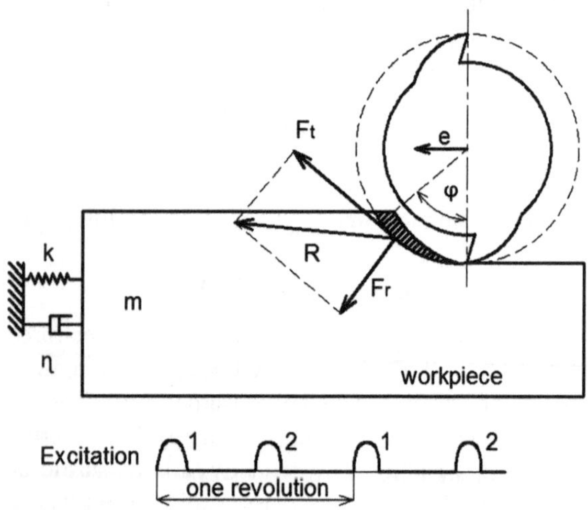

Fig. 5.42 Excitation of a workpiece by periodic cutting forces

The specimen on the machine table is also supported by the elastic rubber which exerts an additional resistance only after a given deformation. Therefore, the displacement curve shows a wavy character.

Softwood species, and particle board, suffer more deformation and absorb more energy. This means less surplus energy and less displacement, which was verified experimentally (Fig. 5.41b).

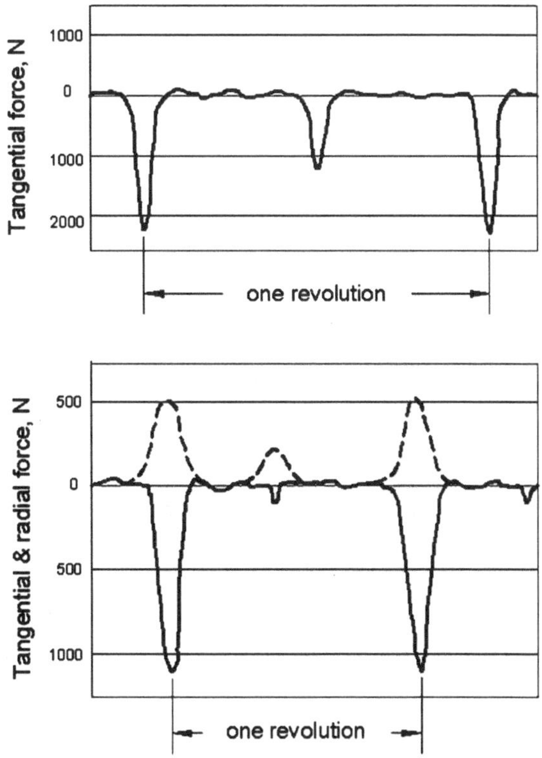

Fig. 5.43 Unequal tooth feed and cutting force due to workpiece vibration clamped on a CNC-router and using a tool with two cutting edges. *Upper Figure*: $n = 3{,}000$ rpm, feed speed 5 m/min, cutting depth 1 mm. *Bottom Figure*: $n = 6{,}000$ rpm, feed speed 5 m/min, cutting depth 1 mm [18]

The pneumatic clamping of a workpiece cannot be regarded as a rigid clamping method and the workpiece is often inclined to vibrate. Especially small workpieces tend to have forced vibrations due to the excitation of the cutting force (Fig. 5.42). The governing equation of motion is similar to Eq. (5.5) with the difference that here also a damping term $(\eta \cdot \dot{y})$ should be included. Resonance occurs if the natural frequency of the workpiece equals the force excitation frequency or its multiple. A typical feature of this phenomenon is the unequal tooth feed and cutting force on the individual cutting edges, as shown in Fig. 5.43.

In the first case the feed per revolution is large enough ($e_n = 1.67$ mm) and the displacement of the workpiece in the same direction to the approaching cutting edge can only halve the tooth feed. In the second case the feed per revolution is 0.83 mm and the second cutting edge has practically no tooth feed. The edge is in contact with the workpiece which is indicated by the radial force. To avoid resonance, the frequency of rotation should be changed. If every second cutting edge fails to cut, the surface will generally show a wavy character similar to washboarding.

Here again an important condition for instability that the tooth passage frequency ($f_t = nz/60$) is slightly higher than the natural frequency of the workpiece. The expected band width of resonance is some 4–5 % related to the resonance frequency of the workpiece.

References

1. Ulsoy, A., Mote, C.: Analysis of bandsaw vibration. J. Wood Sci. **1**, 1–10 (1980)
2. Wu, W., Mote, C.: Analysis of vibration in bandsaw system. For. Prod. J. Sept. **34**(9), 12–21 (1984)
3. Mote, C., Szymani, R.: Principal developments in circular saw vibration. Holz als Roh- und Werkstoff, pp. 189–196 and 219-225 (1977)
4. Лапин, Л.: Определение допустимого числа оборотов круглых пил.(Determination of allowable speed of circular saws.) Лесной Журнал (2), pp. 125–135 (1959)
5. Стахиев, Ю.: Работоспособность плоских круглых пил.(Working capacity of circular saws.) М.: Изд. Лесная Пром. p. 384 с (1989)
6. Стахиев, Ю.: Устойчивость и колебание плоских круглых пил (Stability and vibration of circular saws). Изд. Лесная Пром., (1977)
7. Gogu, G.: Elastic stability and vibration on circular saws. In: Proceedings of 13th IWMS Vancouver, pp. 181–192 (1997)
8. Pahlitsch, G., Friebe E.: Über das Vorspannen von Kreissägeblättern. Holz als Roh- und Werkstoff, pp. 429–436 und 457–463 (1973)
9. Stakhiev, Y.M.: Research on circular saw vibration in Russia: from theory and experiment to the needs of industry. Holz als Roh Werkstoff **56**(2), 131–137 (1998)
10. McKenzie, W.: The effects of slots on critical rim temperature. Wood Science, vol. 4, pp. 304–311 (1973)
11. Satoru, N.: Stable sawblade. In: Proceedings of 17th IWMS Rosenheim, pp. 418–420 (2005)
12. Lister, P et al.: Experimental sawing performance results for super critical speed circular saws. In: Proceedings of 13th IWMS Vancouver, pp. 129–147 (1997)
13. Satoru, N.: Hyper critical sawblade. In: Proceedings of 16th IWMS Matsue, pp. 225–233 (2003)
14. Okay, R et al.: (1995) What is relationship between tooth passage frequency and natural frequency of the bandsaw when wasboarding induced. In: Proceedings of 12th IWMS Kyoto, pp. 267–380 (1995)
15. Lehmann, B., Hutton, S.: The kinematics of wasboarding of bandsaws and circular saws. In: Proceedings of 13th IWMS Vancouver, pp. 205–216 (1997)
16. Orlowski, K., Wasielewski, R.: Washboarding during cutting on frame sawing machines. In: Proceedings of 15th IWMS Los Angeles, pp. 219–228 (2001)
17. Sitkei, G., et al.: Theorie des Spanens von Holz. Fortschrittbericht No.1 Acta Fac. Ligniensis Sopron (1990)
18. Csanády, E., Németh, S.Z.: Investigation of clamping on CNC router. In: Proceedings of 17th IWMS Rosenheim, pp. 456–471 (2005)
19. Dobrindt, P.: Optimiertes Schleifen von MDF—und Spanplatten. In: Workshop Tagungsband, 8.Holztechn. Koll. Braunschweig, S. S. 125–136 (1991)

Chapter 6
The Stability of Wood Working Tools

Processing can only form the expected shape, size and surface quality of a workpiece when the tool works stably and without excessive vibrations. The stability problems, (similar to the general engineering practice), occur in relation to thin and slender tools. These kinds of tools are primarily saws, band saws and circular saws with large diameters.

The main operational and tool design factors influencing cutting accuracy are summarized as follows:

- saw blade dimensions (width, thickness, diameters, free length of the cutting span, tooth pitch and shape)
- preparation of saw blades (roller-tensioning, heating, straining of a band saw, grinding accuracy)
- operational parameters (blade speed, tooth feed, cutting height, cutting depth, wood material properties, tooth sharpness).

Since the number of influencing factors are high, an optimum choice is not a simple task. On the other hand, the operator has any possibilities to improve sawing accuracy and to maximize feed speed.

The loss of the stability of a tool is indicated by the deviation from its original plane (Fig. 6.1). This blade deviation always causes skew cut and the width of the kerf also increases. This can lead in a certain case to the unacceptable decrease of the work-quality and cutting accuracy.

The deflection of the tool from its own plane is caused by eccentric and lateral forces on the cutting edge. The lateral forces especially are dangerous, because saws do not have much lateral stiffness.

The reasons for the development the eccentric and lateral forces on the cutting edge are the geometric irregularities of the succeeding cutting edges, the interaction between the blade's lateral surfaces and the workpiece, band wheel eccentricities and the inhomogeneity of the wood. The slender structural elements deform and buckle easily under compressive stresses. The tangential component of the heat stress developing in circular-saws is always compressive, and this stress component can cause waviness of the disk, the decrease of the natural frequency

E. Csanády and E. Magoss, *Mechanics of Wood Machining*,
DOI: 10.1007/978-3-642-29955-1_6, © Springer-Verlag Berlin Heidelberg 2013

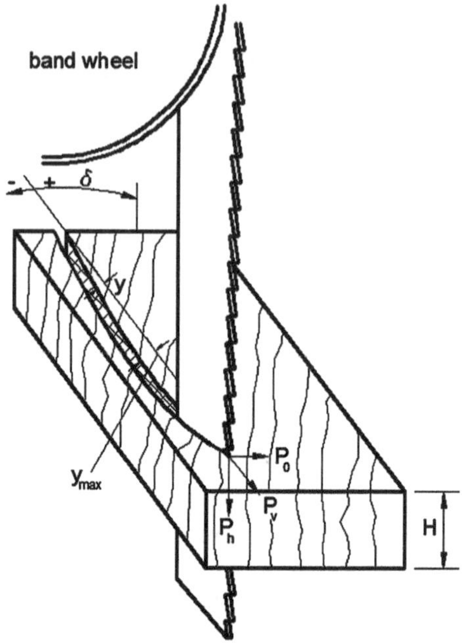

Fig. 6.1 The deflection of the band saw blade from the theoretical cutting plain

and the lateral stiffness (stability). The tangential compressive stress due to heat load can be decreased here with slotting along the periphery and with roll tensioning.

6.1 Cutting Accuracy

The sawn surface always has irregularities due to several disturbing factors such as natural and forced vibrations, lateral forces, non-isotropy of wood, uneven tooth setting etc. These irregularities can be quite different. For example, washboarding (see in Sect. 5.3) is a very regular sinusoidal pattern that sometimes occurs on wood cut by band saws and circular saws. The depth of the corrugation varies from 0.2 to 0.6 mm. This dimensional variation requires the sawyer to cut the lumber thicker than normal to allow the planer to produce a smooth surface.

Uneven tooth setting and the non-isotropy of wood always cause an irregular surface which can be evaluated by statistical methods; (see Fig. 6.2). If the length of the workpiece or time span is long enough for evaluation, then the collected data show a normal distribution and the standard deviation can be determined.

A slight inaccuracy in tooth angles can be detrimental to cutting performance of thinner saws. If the front face is out of square by only 0.5–1.0°, the saw will increasingly lean to one side. The result is a skew cutting with considerable dimension errors. Manual swaging of teeth can often produce an uneven tooth setting with one side heavier than the other. This will also cause skew cutting.

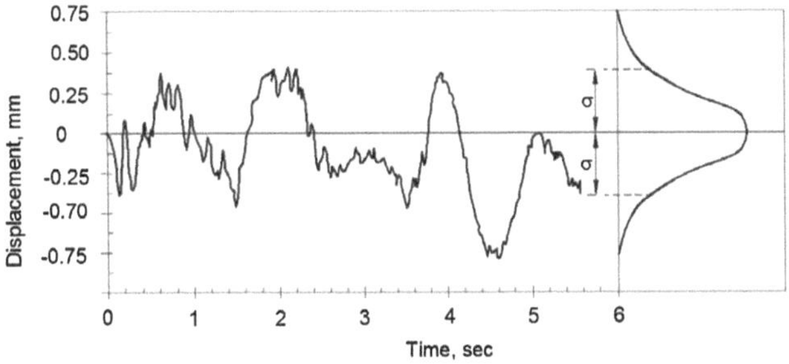

Fig. 6.2 Typical sawing variation profile measured from a sawn cant surface, its normal distribution and standard deviation

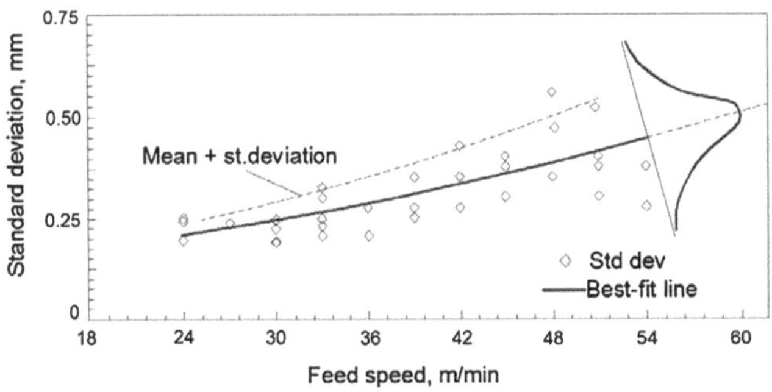

Fig. 6.3 Sawing accuracy vs. feed speed for a circular saw 500 mm in diameter, blade thickness 2 mm, rotation speed 1,900 rpm [1]

Evaluating a set of measurements as a function of a given variable (e.g. feed speed), the standard deviations will be represented in a manner shown in Fig. 6.3 [1].

The points show always a scattering and the line which fits best gives the mean value. Here we can also define the standard deviation of the measured points and it is customary to use the "mean + standard deviation" value. As we see from the Figure, the cutting accuracy decreases at higher feed speeds due to higher forces on the teeth. One of the most important factors influencing sawing accuracy is the lateral stiffness related to the tooth tip or gullet line. Fig. 6.4 shows such a relationship for a band saw with a wheel diameters of 1.5 m, and a band 260 mm wide and 1.5 mm band thick. The feed speed was 60 m/min and blade speed was 48 m/s and with cutting span length of 750 mm [2]. It is apparent from the Figure that a strong relationship exists between lateral edge stiffness and saw blade deviation and, hence, cutting accuracy.

The lateral edge stiffness is a complex parameter influenced by blade diameter or free span length, blade thickness and rolling stresses. The straining of the blade is also a very important factor in band saws.

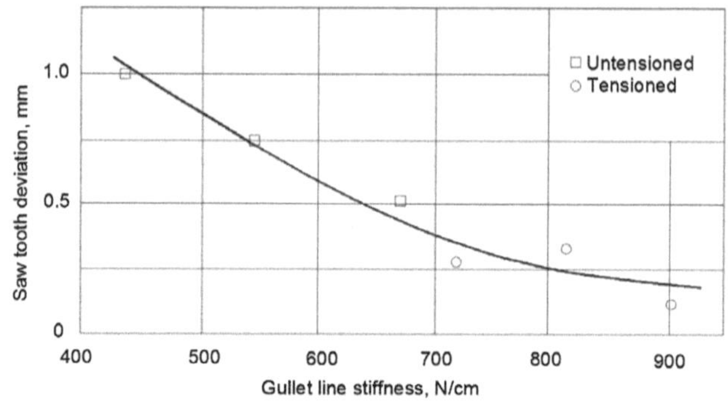

Fig. 6.4 Saw tooth deviation versus gullet line stiffness for a band saw with 1.5 m wheel diameter and 1.5 mm thick blade [2]

In certain cases contacts may occur between the body of the bandsaw blade and the wood. The blade can move laterally in the kerf and if the side clearance (half the difference between the width of the tooth and the width of the blade) is smaller than some critical value, the blade contacts the sawn surface. These contact forces have a significant effect on saw deviation and sawing accuracy. For band mills, the critical value of the side clearance varies around 0.25 mm.

6.2 Blade Tensioning

The tensioning process used in the preparation of saw blades (band saws, circular saws and frame saw blades) has been in existence for about 100 years. The tensioning of blades involves the cold rolling of the blade in longitudinal or circumferential direction between two narrow rollers (Fig. 6.5).

The pressure produced by the rollers causes plastic deformation and permanent residual stresses in the blade material. The art of roller tensioning is to introduce these stresses in a manner that increases the stiffness and natural frequency of the blade and improves the cutting accuracy and feed speed of the saw. Different kinds of stresses may be present in a saw blade, such as initial stresses due to tensioning, rotational stresses and thermal stresses. These stresses lie in the saw blade plane and they are called membrane stresses. During deformation of the blade, these stresses do mechanical work and this work counteracts to a transverse displacement under a given external load. Therefore, an increase in the work of deformation of the stresses stiffens the saw blade. This contribution to the blade stiffness is very important with thin saw blades since the deformation work of the membrane stresses increases relative to the work done by the bending stiffness.

It is well known that thin-walled structures are very sensitive to compression stresses causing buckling. Thermal stresses are especially dangerous because they dramatically decrease the stiffness of the tips of the teeth. These stresses are the

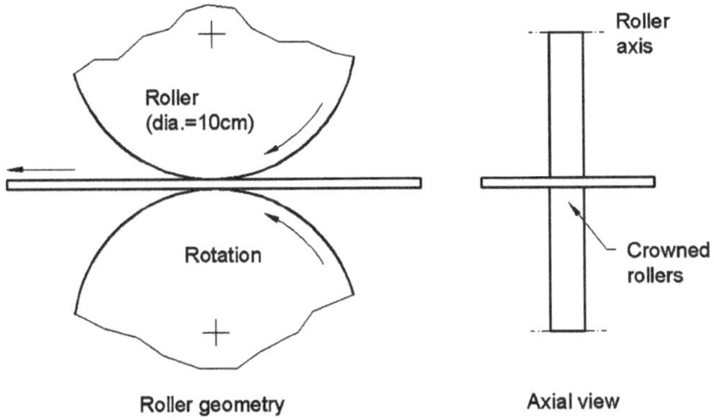

Fig. 6.5 Diagram of the roller tensioning process [2]

dominant cause of saw instability. Increasing the rim temperature decreases the second and higher nodal diameter natural frequencies. As a consequence, the critical speed is reduced. Tensioning should counterbalance the compressive stresses. Therefore, tensioning is particularly significant for thin and large diameter saws where membrane stress effects on saw stiffness are large when compared to the bending stiffness.

Another important factor in band saws is the *blade position on the wheel.* Roll tensioning of the saw blade has been shown to significantly improve the precise positioning of the saw blade on the wheels. A stable blade is desirable in a high production band mill allowing high feed rates with minimal blade movement. Precise positioning of the blade due to tensioning does not require significant amounts of overhang, which have been shown to reduce the stiffness of the cutting edge of the blade. Without tensioning, too little overhang would have the risk of the blade moving back on the wheel and damaging the guides.

The roller crown is the fundamental part of any tensioning machine. During the rolling process, the rolling track should be compressed to establish tensile circumferential stresses around the rim in the tooth zone.

Depending on the shape (profile) of the roller, different $\varepsilon_t/\varepsilon_r$ strain ratios can be achieved, where ε_t is the circumferential strain along the rolling path and ε_r is the radial strain perpendicular to the rolling path. This strain ratio depends on the crown radius or on the width of the roller and on the depth of the rolling track. Their relationship is shown in Fig. 6.6 [3]. In all cases the strain ratio $\varepsilon_t/\varepsilon_r$ decreases with increasing rolling depth.

The rollers generally have diameters around 70–100 mm and a crown radius of 50–100 mm. If a cylindrical surface is used, its width varies between 4 and 6 mm. The rolling position is characterised by the relative rolling radius (r_m/R) for circular saws and by the symmetric distance from the centre line for band- and frame saws [4–6].

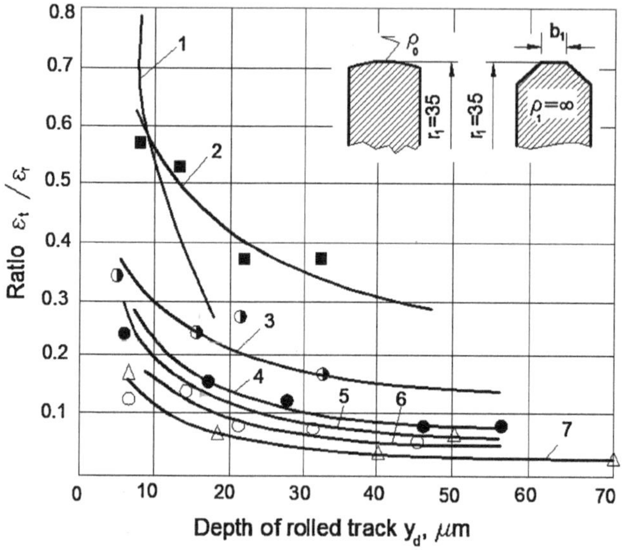

Fig. 6.6 Relationship between the depth of rolled track y_d and the ratio $\varepsilon_t/\varepsilon_r$: (*1*) $\rho_1 = 350$ mm; (*2*) $\rho_1 = \infty$, $b_1 = 10$ mm; (*3*) $\rho_1 = 210$ mm; (*4*) $\rho_1 = 105$ mm; (*5*) $\rho_1 = \infty$, $b_1 = 6$ mm; (*6*) $\rho_1 = 35$ mm; (*7*) $\rho_1 = \infty$, $b_1 = 4.5$ mm [3]

In circular saws, the stresses induced by rolling can be calculated by the following equations:

for the inner part

$$\sigma_{ri} = -\frac{\delta \cdot E}{2 \cdot r_m} \frac{R^2 - r_m^2}{R^2 - r_c^2}\left(1 - \frac{r_c^2}{r^2}\right) \tag{6.1}$$

$$\sigma_{ti} = -\frac{\delta \cdot E}{2 \cdot r_m} \frac{R^2 - r_m^2}{R^2 - r_c^2}\left(1 + \frac{r_c^2}{r^2}\right) \tag{6.2}$$

for the outer part

$$\sigma_{ro} = \frac{\delta \cdot E}{2 \cdot r_m} \frac{r_m^2 - r_c^2}{R^2 - r_c^2}\left(1 - \frac{R^2}{r^2}\right) \tag{6.3}$$

$$\sigma_{to} = \frac{\delta \cdot E}{2 \cdot r_m} \frac{r_m^2 - r_c^2}{R^2 - r_c^2}\left(1 + \frac{R^2}{r^2}\right) \tag{6.4}$$

where R is radius of the blade, r_m is the rolling radius and r_c is the clamping radius. δ is the radial strain due to rolling and its approximate value is calculated as

$$\delta = 3.0\frac{F_r^2}{s^2}\left(0.6\frac{r_m}{R} - 0.1\right) \tag{6.5}$$

Fig. 6.7 Distribution of tangential stresses using different rolling ratios. $D = 500$ mm, clamping ratio 0.3 and rolling force 16 kN [5]

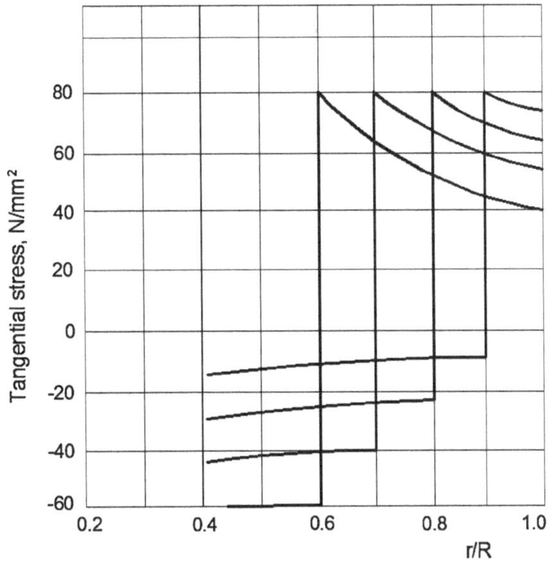

where the rolling force F_r in kN and the plate thickness s in mm must be substituted.

The typical stress distributions for a circular saw 500 mm in diameter are seen in Fig. 6.7 [5]. Tensile stresses occur outside the rolling radius, which stabilize the blade and increase blade stiffness. However, the rim tensile stress slightly decreases with decreasing rolling radius. With an increasing strain ratio $\varepsilon_t/\varepsilon_r$, the stress distribution around the rolling radius will slightly be modified.

There is a maximum rolling force, at which just no static buckling occurs. This load is the *critical rolling force*. The critical rolling force depends on the rolling radius ratio and its value for a given saw diameter and thickness is shown in Fig. 6.8 [3]. In practice, rolling forces slightly less than the critical values are generally used.

Tensioning stresses increase the second and higher nodal diameters natural frequencies. This favourable effect of tensioning stresses depends on the rolling radius. Generally, optimum results can be achieved between rolling ratios of 0.7 and 0.85, as it is shown in Fig. 6.9 [3]. If constant rolling force is used, then the optimum rolling radius is slightly shifted towards smaller rolling radii.

It should be noted that the tensioning stress always decreases the static blade stiffness so the favourable effect of tensioning can be utilized with higher rotation speeds. Figure 6.10 shows the lateral stiffness of a saw blade as a function of rotation speed with and without tensioning [6, 7].

The stiffness of the untensioned blade gradually decreases and tends to zero at the critical speed. The tensioned blade first has an increasing stiffness but after reaching its maximum value, it decreases and tends to zero. The two curves intersect and this

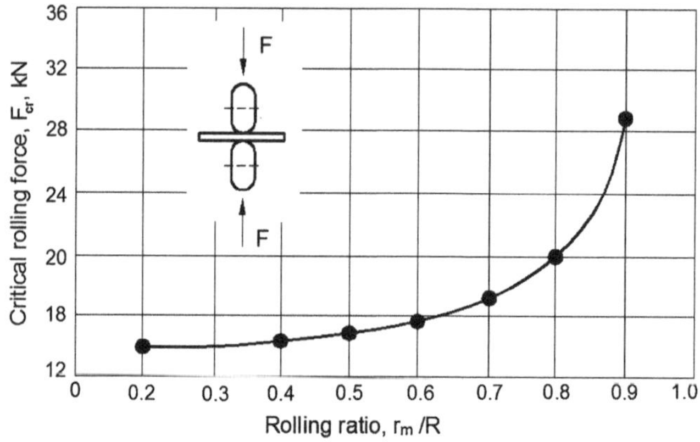

Fig. 6.8 Influence of rolling ratio r_m/R on the critical rolling force F_{cr} corresponding to the mode with zero nodal diameter ($n = 0$) and zero nodal circles ($m = 0$); $D = 500$ mm, $h = 2.2$ mm, $z = 48$ [3]

Fig. 6.9 Determination of the optimum ratio of rolling for rolling along one track: saw diameter = 700 mm, thickness = 3.2 mm, clamping diameter = 160 mm, critical rolling force for each track [3]

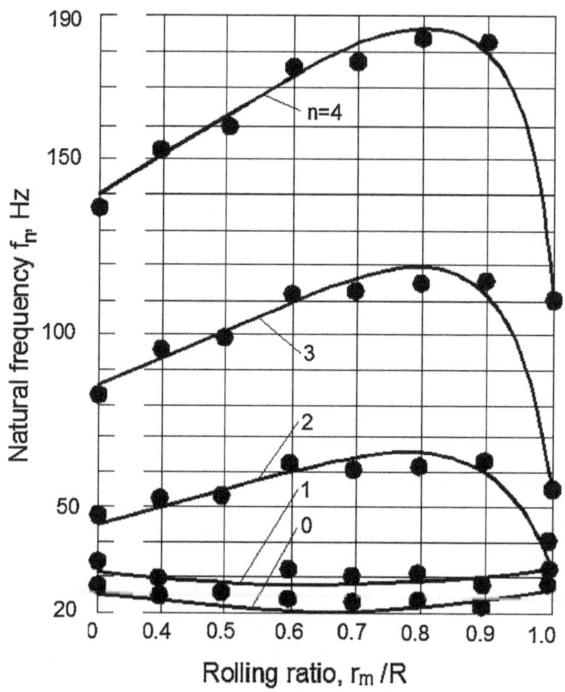

rotation speed is sometimes called "the universal rotation speed." The critical speed increment due to tensioning is generally between 20 and 30 %.

Another method to induce membrane stresses is to use internal pressure acting on the internal circumference of the disk. This has an effect equal to the traditional

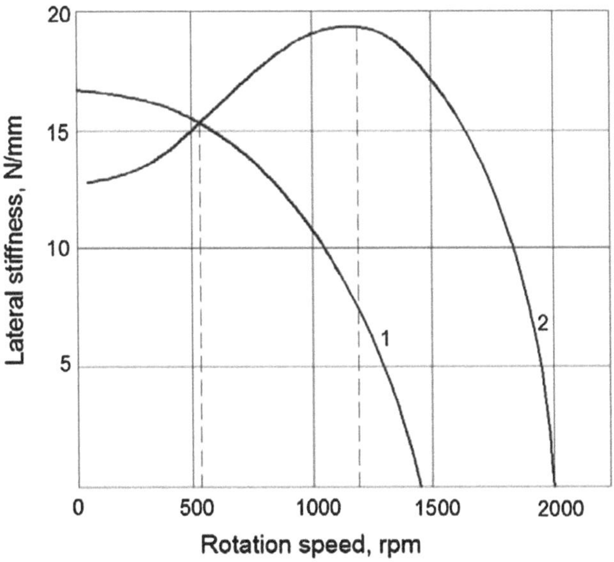

Fig. 6.10 Saw blade stiffness as a function of rotation speed. ($D = 800$ mm, $s = 2.8$ mm) (*1*) without tensioning, (*2*) critically tensioned [6]

roller tensioning method. The in-plane stresses will be obtained by the following equations:

$$\sigma_r = -\frac{r_i^2}{R^2 - r_i^2}\left(1 - \frac{R^2}{r^2}\right) \cdot p_i$$

$$\sigma_t = \frac{r_i^2}{R^2 - r_i^2}\left(1 + \frac{R^2}{r^2}\right) \cdot p_i$$

where r_i means the internal radius of the disc and p_i is the internal pressure. The induced pressure distribution is similar to those of roll tensioning outside the rolling path (Fig. 6.11) [8].

Centripetal tensioning is another tensioning possibility [9]. Centripetal tensioning introduces in-plane residual stresses into the saw blade only while the saw is rotating. This is achieved by a clamp with pivoting wedges that rest against the inner edge of the saw blade, Fig. 6.12 [9]. The advantage of centripetal tensioning is that the residual stresses are only present when there are also centripetal stresses due to rotation. Furthermore, at increasing rotation speeds the centripetal tensioning stresses will also be increased. Theoretical calculations and experimental measurements have verified the effectiveness of centripetal tensioning.

Figure 6.13 shows the lateral stiffness of a circular saw with and without centripetal tensioning. Using this tensioning method, the critical rotation speed can

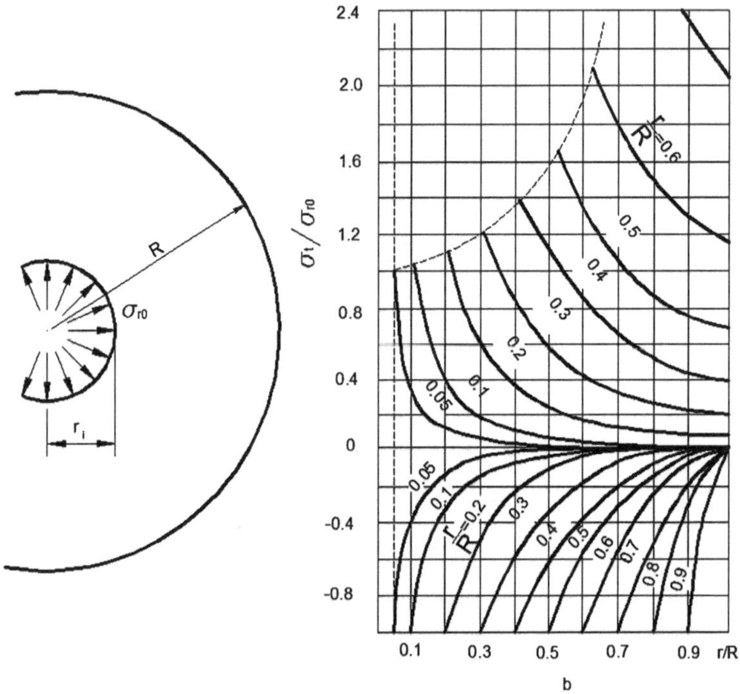

Fig. 6.11 Distribution of tangential and radial stresses due to internal pressure acting on the inner circumference of a circular saw [8]

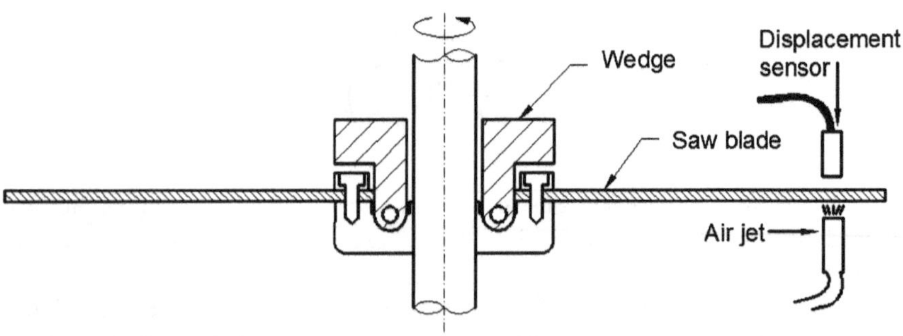

Fig. 6.12 A schematic of a centripetal tensioning clamp and the experimental set up used to measure transverse stiffness [9]

be increased by 20–30 %. At the same time, the design and tensioning mechanism is more complicated and expensive compared to a conventional blade design.

The development of further tensioning methods are in progress [10, 11]. Using high strength steels with case hardening up to 58–60 HRC, the common roll tensioning method is not suitable. A possible solution is the use of a powerful laser

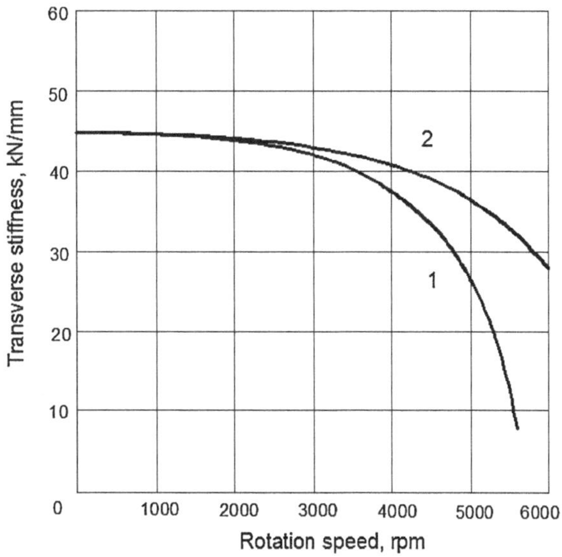

Fig. 6.13 Transverse stiffness versus rotation speed for an untensioned (*1*) and centripetally tensioned (*2*) saw. (Diameter 30.5 cm, blade thickness 1.2 mm, clamping ratio 0.44) [9]

beam (5–6 kW). The laser treatment of steel will introduce residual tensile stresses in the path of the laser so the tensioning path should be positioned at the rim of the blade just below the gullet line.

Another important finding is that the process of manufacturing a saw blade introduces residual stresses, which cannot be left out of consideration. The surface of the saw blade can be ground with cubic boron nitride (CBN) grinding wheels [11, 12]. Grinding of the blade surface causes residual stresses in the near surface layer. They may vary from compressive to tensile values in relation to the grinding depth. The presence of compressive residual stresses in the near surface layer has a damping effect, resulting in smaller vibration amplitudes.

Band saws are also subjected to tensioning to improve blade stability and blade positioning on the wheel. The resultant stress in a band saw is composed of various stress components. The main components are shown in Fig. 6.14 with their distribution along the band width. The conventional roll tensioning occurs along the back line and the gullet line within 2 cm of the rim. Tension stresses occur outside the rolling path while in the central part compression stresses dominate. The induced tensile stresses are around 60 N/mm^2.

The in-plane bending stresses result from wheel tilt, the wheel crown and blade overhang. The third main component is band mill strain having values from 70 to 100 N/mm^2 or even higher.

The degree of tensioning is determined by measuring the deformations of the blade (transverse curvature). In addition, a longitudinal curvature in the plane of the blade may be introduced by rolling near to the back edge. This in-plane curvature is called *backcrown* and its equivalent radius is 700–800 m [2].

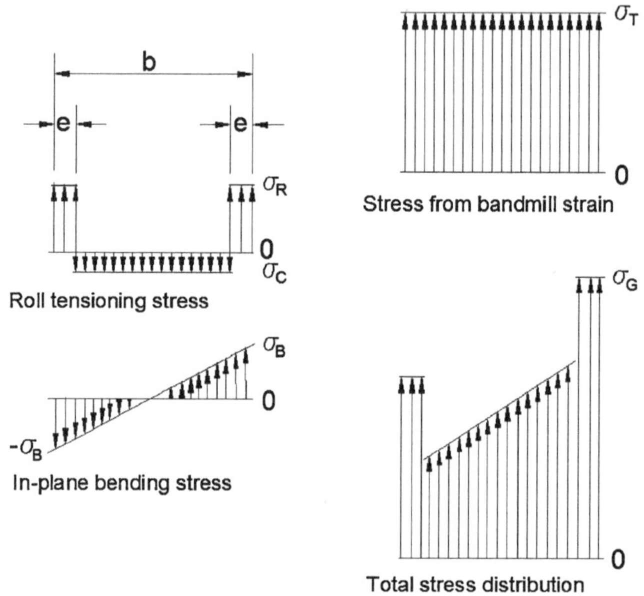

Fig. 6.14 Stress components in a bandsaw

A modified roller tensioning procedure coupled with an increase in band mill strain was also proposed to improve blade stiffness and sawing accuracy over conventionally tensioned blades [2]. The basic idea of this proposal is that keeping the rolled region away from the cutting edge of the bandsaw achieves the most effective tensioning [6]. Therefore, a centre tensioning method is proposed. The two tensioning methods are compared in Fig. 6.15 [2]. Due to the higher distance between the rolling path and the gullet line, the tensile stress induced in the cutting region will be smaller requiring a higher band mill strain. The overall stress level is limited by fatigue life considerations, such as gullet cracking, and this will set an upper limit on the magnitude of the combined stress. The maximum combined stress is generally limited to 150 N/mm^2.

A more detailed rolling profile for centre line tensioning is given in Fig. 6.16 [2]. Rolling occurs in a narrow centre region 40 mm wide. 5 rolling passes in the centre line region are made with 14 kN rolling force and final pair of rolling passes is made at a reduced rolling force of 9 kN, softening the transition from the rolled region to the unrolled region. Centre line rolling ensures a sufficient transverse curvature so that the saw blade is stable on the band mill wheels. The maximum displacement in the centre of saw blade due to the transverse curvature should be around 0.5 mm for a band 260 mm wide and a blade 1.5 mm thick. This value is considered adequate to keep the blade stable on the wheels.

In a running machine, the change of axial band stress due to centripetal acceleration may occur depending on the straining mechanism of the upper wheel.

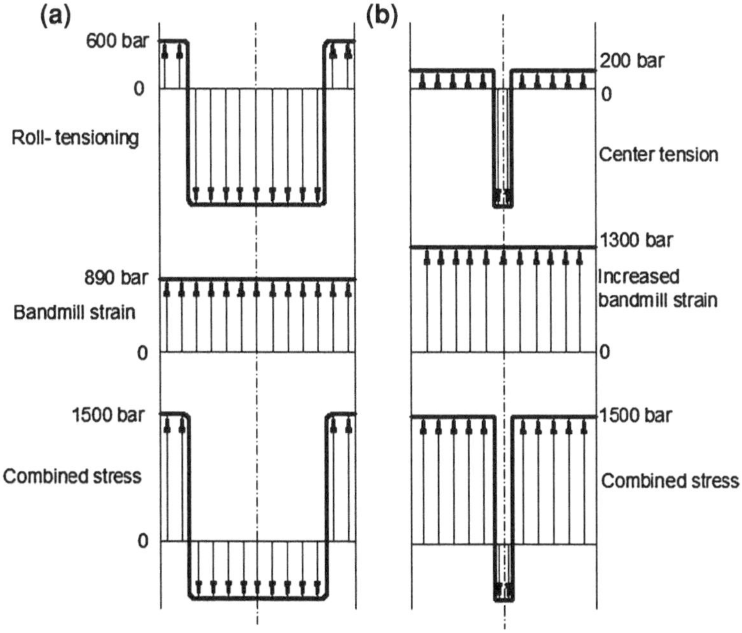

Fig. 6.15 Diagram of stresses for conventional and centre tensioned saws **a** conventional tensioning, **b** centre tensioning [2]

Fig. 6.16 Rolling profile for centre tensioned blades [2]

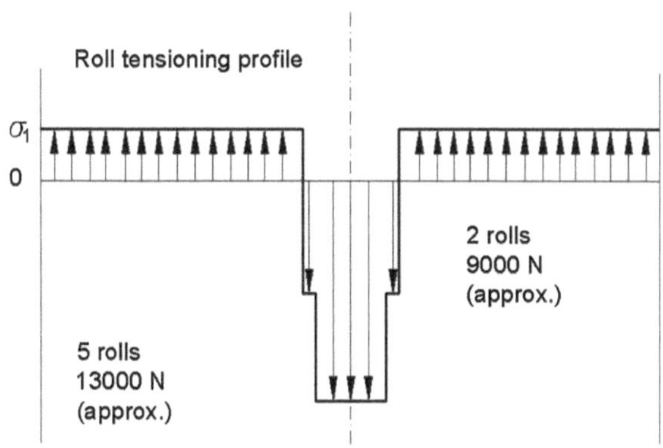

The centripetal force acting on the blade depends on the blade speed and wheel radius

$$F_c = m \cdot \frac{v^2}{R}$$

Integrating the elementary forces along the semi-circular arc we get

$$F_c = \rho \cdot 2b \cdot s \cdot v^2 \tag{6.6}$$

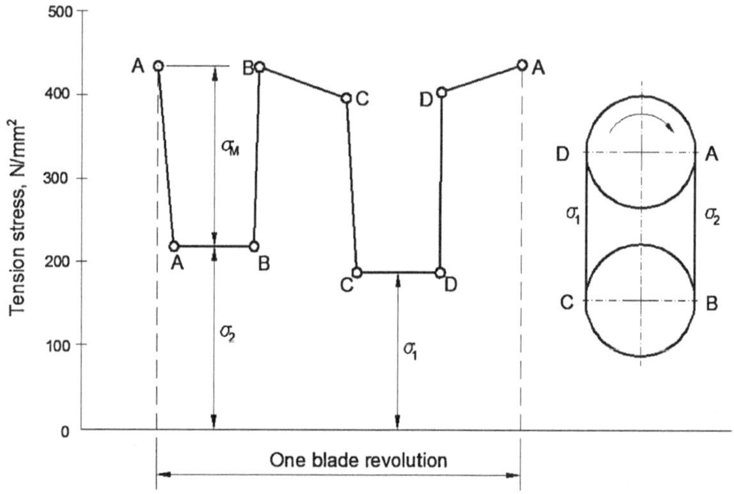

Fig. 6.17 Loading in a band saw blade for one blade revolution

where ρ is the density, b is the width and s is the thickness of the blade.

This force related to the blade cross section yields the centripetal stress

$$\sigma_c = \rho \cdot v^2$$

The centripetal stress decreases the straining stress if its effect is not compensated for by using an appropriate straining mechanism.

The actual stress of a saw blade must be known to determine the fatigue life of a saw blade. The stress loading in the blade varies considerably during one blade revolution due to bending stresses. A blade running on a wheel is subjected to bending and the bending stress can be calculated as

$$\sigma_M = \varepsilon \cdot E = E\frac{s}{D} \tag{6.7}$$

where D means the wheel diameter.

A typical loading cycle for one blade revolution is shown in Fig. 6.17. The loading of the blade is highly cyclic with large amplitude often causing gullet cracking.

The stress in the cutting span is always higher than that of the non-cutting span. These stresses are balanced by the straining and centripetal stresses:

$$\sigma_1 + \sigma_2 = 2 \cdot (\sigma_s + \sigma_c)$$

where σ_S is the straining stress in the blade. The Figure shows the maximum stress in the blade is

$$\sigma_{max} = \sigma_2 + \sigma_M$$

6.3 The Stability of the Band Saw

The forces acting on a single cutting edge of a band saw are given in Fig. 6.1. The P_h cutting force acts in the plane of the blade and along its length, so it can be neglected, as a deflection force.

The P_v feed force may be central or eccentric. The central force does not turn the blade from its plane, but the eccentric feed force is already a deflection force. The most important effect is the lateral force which is also a deflection force.

We have summarized the deformations (deflections), caused by the forces acting on the edge in Fig. 6.18 [13]. The deformation caused by an eccentric force can be put together from two parts. Placing the force into the centre line we get a parallel displacement, to that we add the rotation caused by the $M = P_0 \cdot a$ bending moment (see Fig. 6.18b, c).

The deflection caused by the lateral force can be calculated as follows: The parallel displacement of the blade (Fig. 6.18b) [13]:

$$x = \frac{P_0}{4P_e}\left(L_e - 4\sqrt{\frac{R}{P_e}}\right) \quad \text{and} \quad R = \frac{E \cdot I}{1 - v}$$

the stiffness of the lateral displacement is given by:

$$c_x = \frac{P_0}{x} = \frac{4P_e}{L_e - 4\sqrt{\frac{R}{P_e}}} \tag{6.8}$$

The rotation of a saw blade around the centre of gravity (Fig. 6.18c):

$$b = a \cdot \varphi_0 = \frac{0.75 \cdot P_0 \cdot L_e \cdot a}{a \cdot P_e + 4G \cdot s^3}$$

and the stiffness of rotation:

$$c_\varphi = \frac{P_0}{b} = \frac{4P_e}{3L_e}\left(1 + \frac{4G \cdot s^3}{a \cdot P_e}\right) \tag{6.9}$$

The resultant stiffness is calculated as:

$$c_2 = \frac{P_0}{b} = \frac{c_x \cdot c_\varphi}{c_x + c_\varphi} \tag{6.10}$$

In the previous equations (see also the notations in Fig. 6.18)

P_e pre-stressing force,
L_e the free length of the band without clamping,
G the modulus of elasticity in shear,
s the thickness of the band.

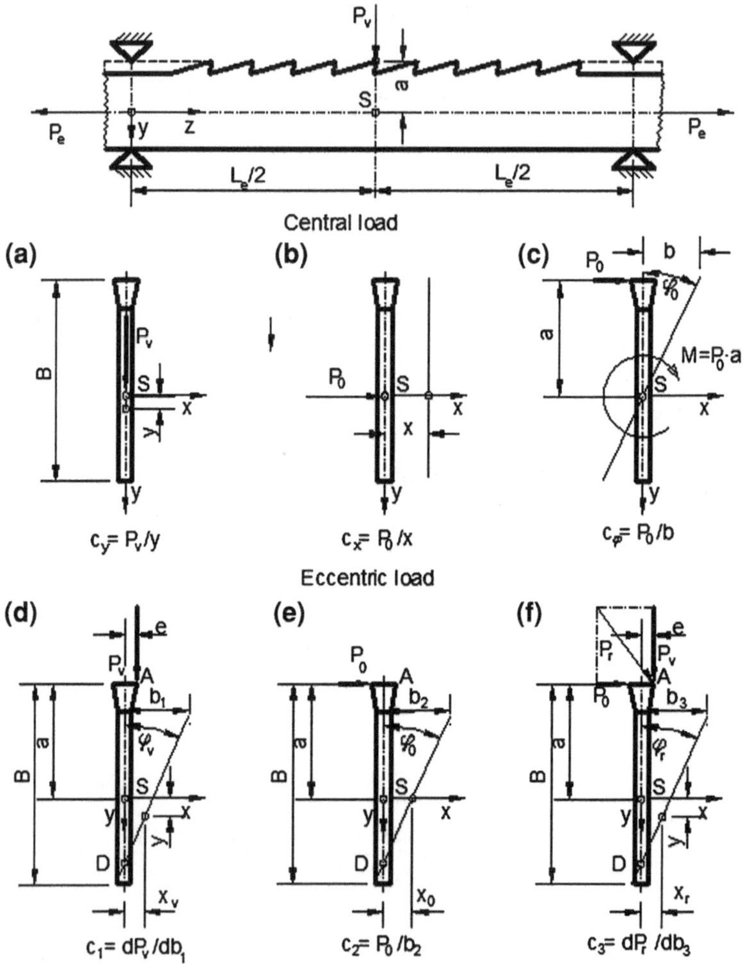

Fig. 6.18 Different possible forces acting on the edge of a band saw and the definition of appropriate stiffness coefficients [13]

A greater stiffness causes better stability of the band. The stiffness is mostly determined by the pre-stressing force and the free length of the blade. Therefore on a band saw, it is important to use a band guide and to set it correctly. The width and thickness of the band has less influence on its stability.

The stiffness has remained constant depending as a function of displacement. However, eccentric feed forces (Fig. 6.18d, f), cause the most stiffness at the beginning of the deflection. As deflection increases, the stiffness decreases. Stiffness also varies from the combined effect of eccentric feed force and the lateral force. Stiffness decreases dramatically as the lateral force increases. Their relationships are shown in Figs. 6.19 and 6.20 [13].

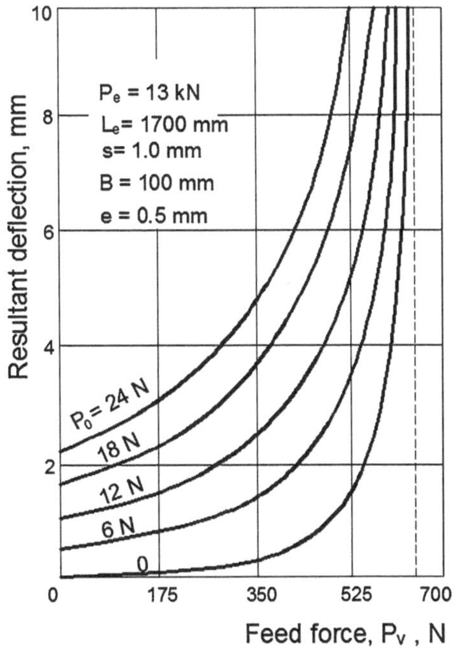

Fig. 6.19 The resultant deflection of a band saw depending on the feed force for different lateral forces [14]

With increasing feed forces, deflection rises. After reaching a critical value, the band turns 90° away from its original plane. At the same time the stiffness decreases to zero (Fig. 6.19 and 6.20). The critical feed force can be calculated from the following equation [13]:

$$P_{vcr} = -\frac{a \cdot c_x}{2k^2} + \sqrt{\frac{a^2 \cdot c_x^2}{4k^2} + \frac{a^2 \cdot c_x \cdot c_\varphi}{k^2}} \quad (6.11)$$

where:

$$k = \frac{4 + \pi^2}{16} = 0.867$$

Band saw blades are subjected to roller tensioning to improve blade stability and blade positioning on the wheel. Roller tensioning stresses (Fig. 6.14 and 6.15) considerably increase the lateral blade stiffness and decrease the torsional vibration amplitudes.

Since there is a strong relationship between the lateral edge stiffness and cutting accuracy, the resultant stress distribution in the blade should provide the highest possible lateral edge stiffness; (see Fig. 6.4).

In practical cases, the feed speed is an important factor determining sawing performance. An increasing feed speed always increases the feed force (radial cutting force) and its lateral components causing higher tooth tip deflection. Both

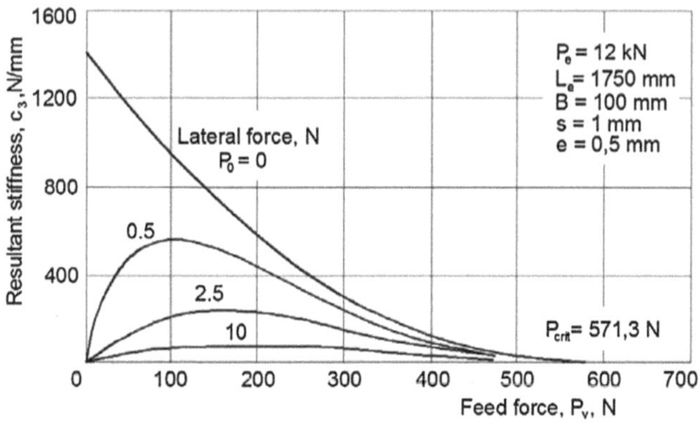

Fig. 6.20 The resultant stiffness of a band saw depending on the feed force [14]

tooth tip deflection and sawing accuracy are strongly related to the feed speed; (see Fig. 6.3).

Based on theoretical considerations and practical observations, the following main conclusions may be drawn:

- a precise and careful tool preparation (spring or swage setting, grinding accuracy) highly decreases the presence of eccentric and lateral forces, which would increase tooth tip deflection;
- eccentric and lateral forces due to wood failures and inhomogeneity of the timber can be decreased primarily by decreasing of the feed per tooth.
- the deflection of the saw blade from its plane increases linearly under the effect of the increase of the feed force;
- the increase of the pre-stressing increases the value of the critical feed force, and improves the stability;
- the free span length of the saw blade must be as small as possible, as the band guide prevents the rotation of the blade;
- introduction of tensioning stresses in a manner that increases the lateral stiffness of the blade in a favourable way and improves the cutting accuracy of the band saw.

6.4 Band Saw Tracking Stability

The sideways motion of belt pulleys can be avoided by using crowned wheels [15, 16]. A saw blade running on the band mill wheels is subjected to feed forces, which tend to move the saw blade "front-to-back." This sideways motion of the saw blade is called *tracking*. The ability of a moving saw blade to counteract the cutting forces and to maintain its initial and stable position on the wheel is referred to as *tracking stability* [16, 17].

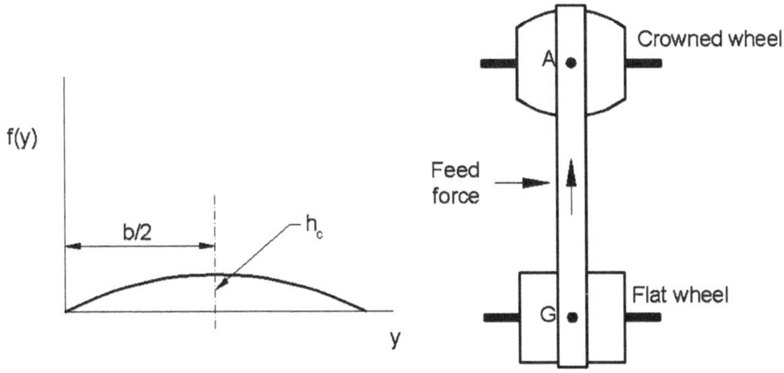

Fig. 6.21 Crowned band mill wheel and its profile [17]

During sawing, the feed forces (cutting forces) create an in-plane bending moment and curvature along the free length of the saw blade between the wheels. As a consequence, the saw blade approaches the wheel at a small angle and moves onto the wheel with a screw-like motion. This causes a sideways motion of the saw blade on the wheel.

To maintain the saw blade in its correct position on the wheel, the band mill wheel should generate a bending moment to counteract the moment generated by the feed forces. This can be achieved by using an appropriate wheel profile and saw blade overhang. The balance between the two moments determines the displacement of the saw blade and it can keep the saw blade on the wheel in its correct position. Having an appropriate tracking stability, the saw blade will quickly return to its initial position after any sideways displacement caused by a disturbance, such as a feed force. The saw blade moves only a small distance in response to the feed force.

Figure 6.21 shows a crowned band mill wheel and its circle profile. The profile can easily be described in the following manner:

$$f(y) = h_c \cdot \left(1 - \left(\frac{y - \frac{b}{2}}{\frac{b}{2}}\right)^2\right)$$

where b is the width of the wheel and h_c is the crown height. The crown height is generally two per thousand of the width of the wheel.

There are different ways to achieve appropriate band saw stability. Figure 6.22 shows several possibilities, such as a crowned wheel, a tapered wheel and band overhang. Tilting a flat wheel corresponds to a tapered wheel.

The feed force (cutting force) generates a bending moment in the band. Therefore, the band between the wheels behaves as a beam bending under an axial tensile load with a centre force acting in the span. The bending moment at any point along the length of the band is

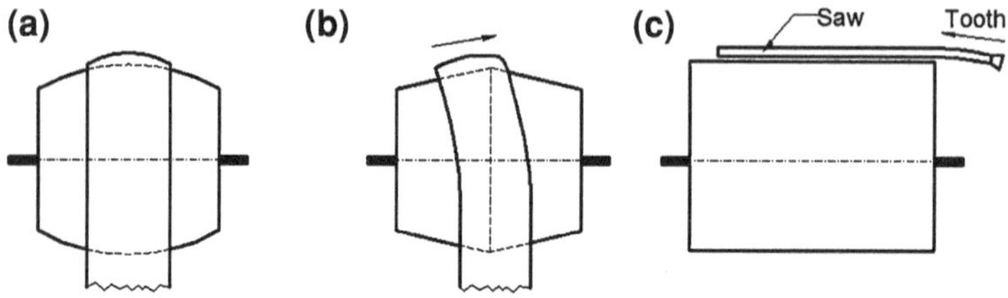

Fig. 6.22 Crowned wheel (**a**), double tapered wheel (**b**) and the tracking effect of overhang (**c**) [17]

$$M_1 = E \cdot I \frac{d^2 y}{dx^2}$$

where $E \cdot I$ is the flexural rigidity of the band and $\frac{d^2 y}{dx^2}$ gives the band curvature.

The mechanism of band saw tracking stability can be examined on a tapered wheel shown in Fig. 6.23 [17]. The tapered wheel applies the strain unevenly to the band, which results in a bending moment and consequently a curvature in the band. The first contact point **A** on the wheel tends to follow a path perpendicular to the rotational axis to point **C**. Therefore, the band moves sideways towards the wider side of the tapered wheel.

To understand the behaviour of a crowned wheel, it may be replaced by a double tapered wheel; (see Fig. 6.22b). Since the band is running on the left tapered side, this side tends to move the band towards the centre line. As the band approaches the centre of the wheel, the influences of the left and right taper sides will be equalized and the band will reach an equilibrium position. Similarly, the band running on a crowned wheel tends to remain at the top of the crown. That means, the crowned wheel centres itself. In Fig. 6.22c, we see a band overhanging the flat wheel that curls down towards the wheel axis. The curling of the band has a similar effect as that of the tapered wheel and creates also an uneven stress distribution, and a bending moment in the band. Therefore, the use of a band overhang on a flat wheel will also centre itself.

To counterbalance the bending moment of the feed force, the band should be running slightly asymmetrically on the crowned wheel, ensuring the appropriate bending moment in opposite direction. With increasing feed force, the asymmetry of the band on the wheel increases linearly.

Tilting the wheel sets the band overhang. In the vicinity of a zero tilt angle, the overhang is small and the position of the band is very sensitive to small tilt changes. Over a 0.1° tilt angle, the relationship between overhang and tilt angle is more or less linear; the overhang increases with wheel tilt.

The tracking stability of the band can be increased by combining the overhang and wheel crown. The band will not move sideways as much in response to a feed force.

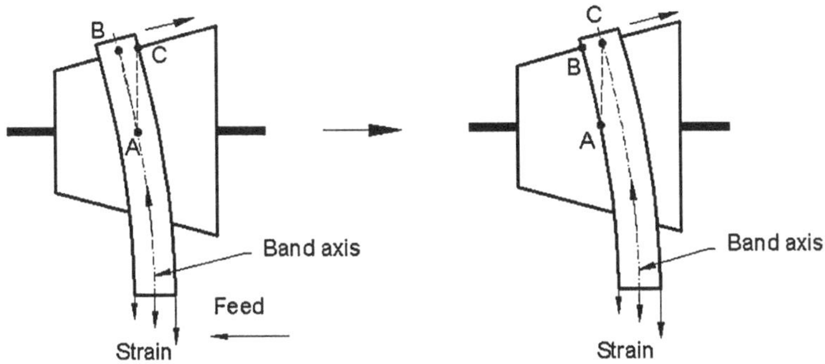

Fig. 6.23 Band motion on a tapered wheel, initial position and position after ¼ turn [17]

6.5 Stability of Circular Saws

Saws operating in different conditions always vibrate, but this does not necessarily mean instability. A sign for impending transverse instability is when any small disturbance causes large transverse displacements of the saw from its current equilibrium.

Two different instability mechanisms have been observed in saw vibration. The first mechanism is the static buckling, where the saw blade has a constant harmonic shape with large amplitudes. This mode of buckling consists of low nodal diameters (from 0 to 1 nodal diameter) and the blade often rubs against the workpiece. The latter action is associated with intensive frictional heating which causes more instability of the blade.

The second instability mode is the critical speed instability, (see in Sect. 5.2), where the backward wave frequency equals the rotational speed frequency causing a standing wave; (see Fig. 5.9). The critical rotation speed determines the maximum stable rotation speed of the saw which in the practice is 15 % less than the critical speed.

The critical speed instability always occurs before buckling and is the more significant and important instability mode.

Saw vibration has many drawbacks: increasing the kerf loss and energy requirements, reducing cutting and product accuracy, diminishing surface quality etc. Therefore, efforts should be made to reduce vibration amplitudes to a low level. This can be accomplished through appropriate saw design and membrane stress modification.

The vibration amplitudes strongly depend on the lateral stiffness of the saw blade. The static lateral stiffness can be calculated in the following manner [8]:

$$I_{st} = \frac{F}{y_{st}} = \frac{E \cdot s^3 \sqrt{R^2 - r_c^2}}{25(R - r_c)^3} \quad \text{N/m} \tag{6.12}$$

where E is the modulus of elasticity, N/cm^2, s is the blade thickness, cm, R is the outer radius of the blade, cm, and r_c is the clamping radius, cm.

The above equation shows that the blade thickness has a very dominant influence on its lateral stiffness. In big diameters saws, it is important to introduce appropriate membrane stresses to increase their natural frequency and lateral stiffness; (see in Sect. 5.2).

As shown in Fig. 5.10, different membrane stresses occur in a rotating disk influencing its vibration mode, critical speed and also the stability of the saw blade.

The rotation of the saw induces only tensile stresses, which increase the stability of the saw. This can also be seen in Fig. 5.9, where natural frequency of the saw increases monotonically with its rotation speed. However, if the blade has unbalanced masses, then this moving load can cause instability when there is moving load resonance at critical speed.

A more important factor is the presence of thermal stresses. The magnitude of thermal stresses depends on the coefficient of thermal expansion of the blade material and the temperature distribution in the blade induced by friction forces. The thermal stresses in a tangential direction are always compressive and they are the dominant cause of saw instability. An increasing rim temperature decreases the natural frequencies of saw blades with nodal diameters of two or more, consequently reducing the critical speed. The saw becomes less stable at lower rotation speeds. Different measures can be used to counteract the unfavourable effect of thermal stresses. One way is to use an induction heater to heat the central region of the blade and produce thermal tensioning. This modifies the temperature distribution of the blade and lowers the temperature difference between the central and outer regions of the blade. Thermal stress depends on the temperature difference but not on the temperature itself.

Another frequently used measure is the use of slots along the periphery of the blade. The slots reduce compressive hoop stresses, allowing free expansion of the periphery without developing tangential stresses. The slots have a very complicated effect on blade stiffness, natural frequencies and vibration amplitudes. Therefore, contradictory experimental results were often obtained with slotted saws.

An interesting phenomenon associated with slots is that the number of resonant frequencies is doubled because of the frequency splitting due to the slots. At the same time, the excitation energy is more distributed throughout the spectrum, reducing the vibration amplitude compared to a saw without slots.

Tensioning, as discussed in detail in Sect. 6.2, is an artificially introduced residual stress in the saw through plastic deformation by rolling and/or through local heating. Optimum tensioning means the introduction of a favourable stress state, increasing the natural frequencies and the critical speed of the saw blade.

The main practical problem is that we do not have accurate non-destructive measurement methods to determine initial stresses in a saw blade. Therefore, two indirect methods of residual stress evaluation are generally used. Either we measure the natural frequencies of the saw and critical speed or its stiffness. The results of both measurement methods generally correlate to one another, so stiffness can be regarded as a measure of tensioning effect on the natural frequencies of a saw and its critical speed. The lateral stiffness of a saw can be measured under static and dynamic conditions. A dynamic measuring method is needed to obtain the stiffness characteristics as a function of rotation speed; (see Fig. 6.12).

Thermally induced stresses are highly dependent on the operational parameters, such as the cutting height, feed speed and rotation speed. Therefore, the full compensation for these stresses by a constant tensioning stress is not possible. Attempts have been made to overcome this problem through continuous adjustment of the state of stress using on-line saw stability control [8]. Temperature differences along the blade radius can be eliminated by either local heating or cooling (Fig. 6.24).

Using temperature or displacement sensors, the duration or intensity of heating/cooling can easily be regulated. The displacement sensor measures vibration amplitudes, which indirectly refer to increasing temperature differences. Figure 6.25 shows the temperature course of a saw blade without and with water cooling during sawing. Appropriate cooling can practically eliminate temperature differences in a saw blade.

The results in Fig. 6.25 make it possible to draw some conclusions regarding thermal inertia of saw blades. In the transition from one operating condition to another the saw blade exhibits a thermal inertia, which depends on the geometrical dimensions of the blade and on its temperature. That means, under varying operational conditions warming-up and cooling-down periods occur.

After a sudden load increase (step function), the temperature in the blade varies with the following equation:

$$\vartheta = \vartheta_0 + (\Delta\vartheta)_{\max}\left(1 - e^{-\frac{t}{T}}\right) \tag{6.13}$$

where ϑ_0 is the ambient temperature and $(\Delta\vartheta)_{max}$ is the maximum temperature rise until a steady-state condition. The time constant T characterizes the speed of temperature change following a load change. More exactly, the value of T indicates how much time is required to reach 63 % of the maximum temperature rise $(1 - 1/e)$. This value is $T \approx 30\ s$ slightly increasing towards the inner parts due to the decreasing temperature.

Generally we should not reckon with a creep relaxation of the original stress state due to roller-tensioning. That means that a saw should be rolled only once and this initial stress state with cyclic loading remains the same over time and it shows no tension decay.

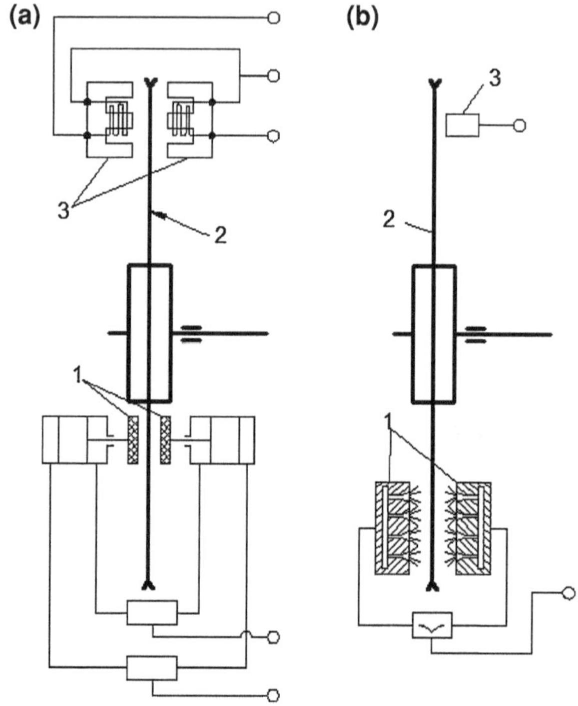

Fig. 6.24 On-line stability control of a saw blade using heating (**a**) or cooling (**b**). **a** heating with rubbing pads, **b** cooling with water jets [8]

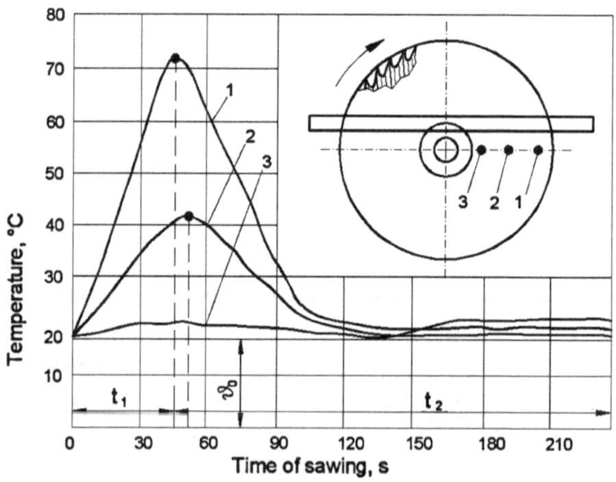

Fig. 6.25 Temperature course of the blade without cooling (t_1) and with continuous water cooling (t_2)

6.6 Frame Saw Stability

Frame sawing machines in many European countries rank among the basic equipment for primary wood processing. It is interesting to note that the first ancient saw mill used in the 15 and 16th centuries was a man-driven frame saw,

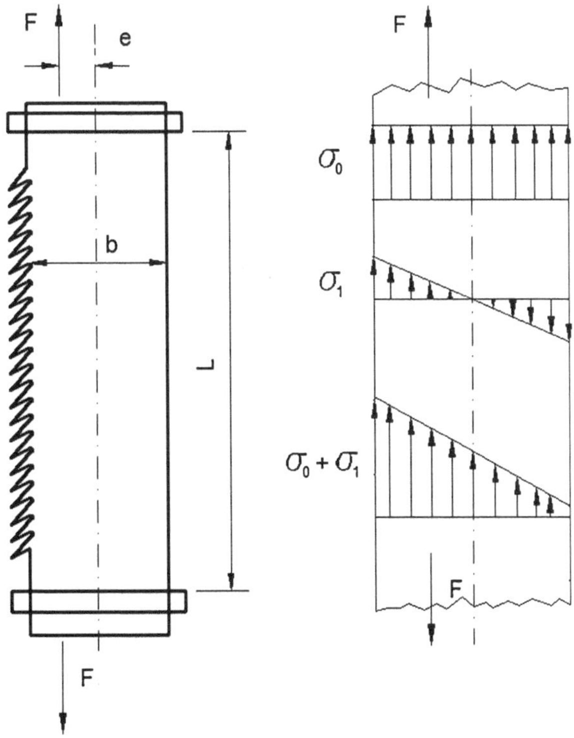

Fig. 6.26 Stresses in the blade of a frame saw

which was first depicted in Bessoni's "Theatrum Instrumentorum et Machinarum" in 1578.

Similarly to other sawing machines, the stiffness and stability of the saw blade determine sawing accuracy, surface roughness and the occurrence of washboarding. The saw blade can be regarded as a beam with a rectangular cross section with a free length L, blade width b and blade thickness s (Fig. 6.26). Both ends of the blade are clamped and strained with a force F. Generally the blade is strained with an eccentricity e creating a more favourable stress distribution in the blade.

The straining stress σ_0 is calculated as

$$\sigma_0 = \frac{F}{b \cdot s} \tag{6.14}$$

which generally amounts to between 100 and 200 N/mm^2.

Eccentric loading generates a pure bending moment and, as a consequence, a bending stress

$$\sigma_1 = \frac{F \cdot e}{I} \cdot \frac{b}{2} \quad \text{with} \quad I = \frac{s \cdot b^3}{12}$$

and

$$\sigma_1 = \frac{6F \cdot e}{s \cdot b^2} \tag{6.15}$$

The resultant stress at the gullet line is

$$\sigma_g = \sigma_0 + \sigma_1$$

while along the back line it amounts to

$$\sigma_b = \sigma_0 - \sigma_1$$

This latter resultant stress may not be negative, which would mean a compressive stress causing instability.

The neutral axis will be shifted to the right side with a value of

$$n_a = \frac{b}{2}\frac{\sigma_0}{\sigma_1} = \frac{b^2}{12e} \tag{6.16}$$

If $e = b/6$, then the above equation yields

$$n_a = \frac{b}{2}$$

which means that the neutral axis lies on the back rim of the blade having a zero resultant stress.

The elongation of the blade under strain forces amounts to a measurable quantity:

$$\Delta L = \frac{\sigma_0}{E} L$$

which can be used to control the expected strain force.

During woodworking operations the blade temperature increases, which causes elongation of the blade

$$\Delta L_t = \alpha \cdot L \cdot \Delta\vartheta$$

where α is the thermal expansion coefficient (for steel $\alpha = 12 \cdot 10^{-6}$), and $\Delta\vartheta$ is the temperature difference. The elongation of a saw blade decreases the initial strain stress σ_0 with the following value

$$\Delta\sigma = -\alpha \cdot E \cdot \Delta\vartheta$$

For example, setting a strain stress $\sigma_0 = 150$ N/mm^2 and the temperature rise is $\Delta\vartheta = 60\,°\text{C}$, then $\Delta\sigma \approx 150$ N/mm^2 entirely stopping the strain on the blade.

The most common vibration mode of a saw blade is lateral although torsional vibration can also occur. The possible lateral vibration modes are shown in Fig. 6.27.

Due to the clamping of the blade at both ends, the end points are always nodal points. The natural frequencies can be calculated from the following equation [18]:

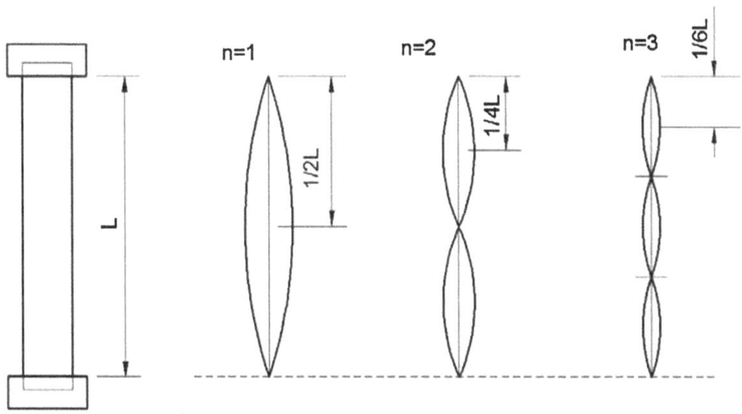

Fig. 6.27 The first three vibration modes of a blade clamped on both ends

$$f_n = \frac{n}{2L}\sqrt{\frac{n^2\pi^2}{L^2}\frac{E \cdot I}{\rho \cdot b \cdot s} + \frac{F}{\rho \cdot b \cdot s}} \tag{6.17}$$

where ρ is the density of the blade material and $n = 1, 2, 3$ denotes the number of vibration modes. The flexural rigidity of the blade $E \cdot I$ is given as

$$E \cdot I = E\frac{b \cdot s^3}{12}$$

Figure 6.28 shows the first natural frequencies for a given blade as a function of strain force (or strain stress). The relationship is not linear and at zero strain, the blade has a low natural frequency, only 3.3 Hz.

The main component of the cutting force lies in the saw blade plane and it causes no lateral deflection. However, the cutting force always has smaller or larger lateral component resulting in lateral displacement of the tooth tip. The tooth tip deflection will be determined by the lateral stiffness of the blade and its lateral forces.

The behaviour of a frame saw blade to lateral forces is quite similar to that of a band saw blade; (see in Sect. 6.3). The approximate calculation method given for band saw blade can also be used for frame saw blade. The behaviour of a frame saw blade subjected to lateral forces is given in Fig. 6.29. The deflection of the tips of the saw teeth has two components due to the blade lateral motion and blade rotation. Each component, u_0 and u_1, can be calculated and also the resultant stiffness coefficient can be determined.

The strain force has a fundamental influence on both the tooth tip deflection and stiffness. This relationship for a given blade is given in Fig. 6.30. The tooth tip deflection is calculated for a lateral force of 10 N. The tooth height h is defined as the distance between gullet line and tooth tip.

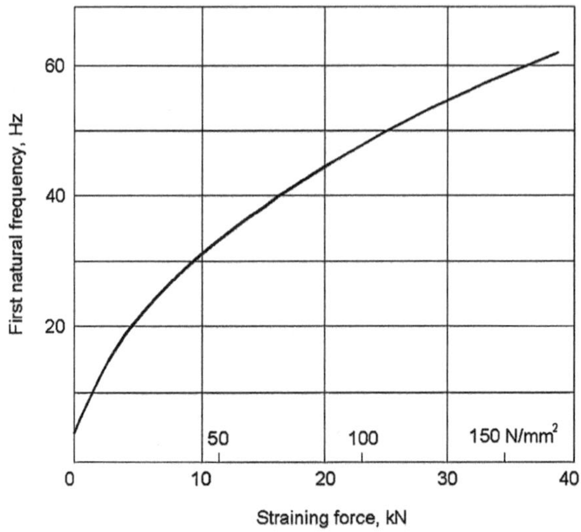

Fig. 6.28 First natural frequency versus strain force relationship. $L = 1.2$ m, $b = 115$ mm, $s = 2$ mm

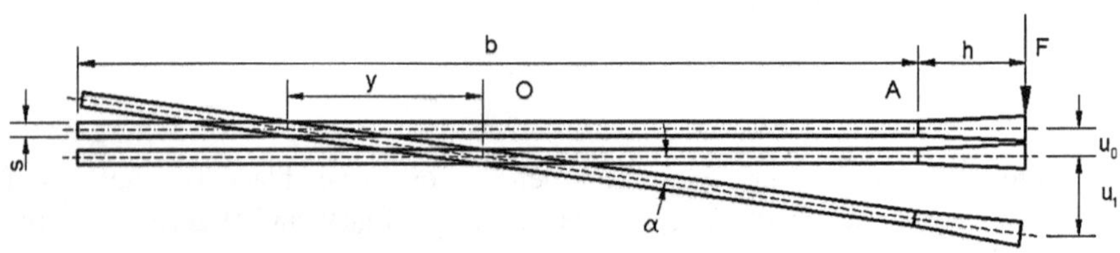

Fig. 6.29 Tooth tip deflection of a saw blade subjected to lateral load

The actual point of rotation of the blade is always behind the mid-point of the blade (see Fig. 6.29) with a distance of

$$y = \frac{u_0}{u_1}\left(\frac{b}{2} + h\right) \tag{6.18}$$

In some cases washboarding was also observed in frame saws using mini saw blades [19]. Due to the reciprocating motion of the frame saw, the tooth passage frequency is not constant but varies within one cycle according to the following expression

$$f_t = \frac{\pi \cdot f_r \cdot H \cdot \sin\varphi}{p} \tag{6.19}$$

where f_r is the frequency of rotation, H is the frame stroke and p is the tooth pitch.

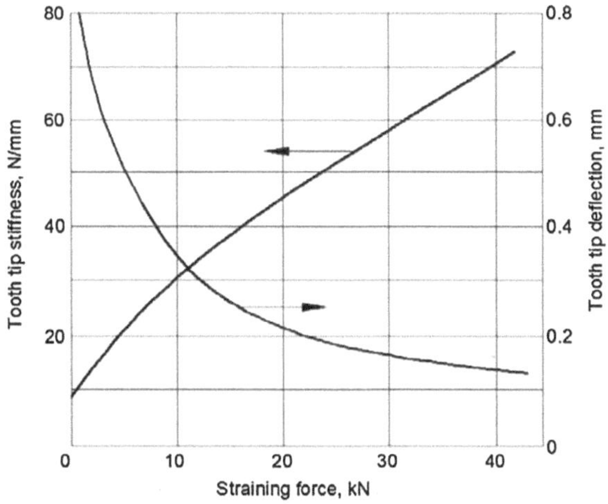

Fig. 6.30 Tooth tip stiffness and deflection as a function of the strain force. $L = 1$ m, b = 120 mm, $s = 2.2$ mm, $h = 15$ mm

To use a constant value, the root mean square tooth passage frequency may be used, which is roughly 70 % of the maximum value calculated from Eq. (6.19). It is supposed that the cyclic washboarding profile is generated more by forced vibration of the blade due to lateral forces. However, if the ratio of the mean tooth passage frequency and natural frequency of the blade is close to one, then the height of waviness increases.

Frame saw blade can also be roll-tensioned to improve the stiffness of the tooth region using three symmetrically spaced roller passes [6].

References

1. Lister, P. et al.: Experimental Sawing Performance Results. In: Proceedings of the 13th IWMS Vancouver, pp. 129–147 (1997)
2. Taylor, J., Hutton, S.: Simplified bandsaw tensioning procedure for improve blade stiffness and sawing accuracy. For. Prod. J., pp. 38–44 (1995)
3. Stakhiev, Yu.: Research on Circular Saw Tensioning in Russia. In: Proceedings of 15th IWMS Los Angeles, pp. 293–302 (2001)
4. Стахиев, Ю., Определение оптимального расположения ширин зоны вальцевания дисков пил (Optimum rolling radius and width for circular saws). Изд. Лесная Пром. с. 23–25 (1966)
5. Pahlitsch, G., Friebe, E.: Über das Vorspannen von Kreissägeblättern. Holz als Roh- und Werkstoff, S. 457–463 und S. 5–12 (1974)
6. Bajkowski, J.: Spannugnsverteilung in durch Walzen vorgespannten Gattersägeblättern. Holztechnologie, S. 258–262 (1967)
7. Stakhiev, Y.-M.: Today and tomorrow circular sawblades: russian version. Holz als Roh- und Werkstoff **58**, 229–240 (2000)
8. Санев, В.: Обработка древесины круглими пилами.(Woodworking with circular saws.) Изд. Лесная Пром. (1980)
9. Renshaw, A., Centripetal Tensioning for High Speed Circular Saws. In: Proceedings of 14th IWMS Paris, pp. 129–135 (1999)

10. Münz, U., Thiessen, B.: Straightening and Tensioning of High Hardness Circular Saw Blades by Laser Beam. In: Proceedings of 16th IWMS, Matsue, pp. 234–247 (2003)
11. Münz, U.: Vibration Behaviour and Residual Manufacturing Stresses of Circular Saw Blades. In: Proceedings of 17th IWMS, Rosenheim, pp. 407–417 (2005)
12. Маталин А.А., Причины возникновения остаточных напряжений. В кн. » Качество обработанных поверхностей » .(Origin of residual stresses. In:Quality of machined surfaces.) Лонитомаш. Кн. 34 М.—Л. Машгиз (1954)
13. Pahlitsch, G., Putkammer, K.: Beurteilung für die Auslenkung von Bandsägeblättern. Holz als Roh- und Werkstoff, S. 295–302 (1974)
14. Pahlitsch, G., Putkammer, K.: Beurteilung für die Auslenkung von Bandsägeblättern. Holz als Roh- und Werkstoff, S. 413–426 (1976)
15. Sugihara, H.: Theory of Running Stability of Band Saw Blades. In: Proceedings 5th Wood Mach. Sem. UC For. Prod. Lab. Rich., CA, pp. 99–110 (1977)
16. Swift, H.W.: Cambers for Belt Pulley. In: Proceedings Institution of Mechanical Engineers, London, vol. 122, pp. 627–659 (1932)
17. Wong, D., Schajer, G.: Effect of Wheel Profile in Bandsaw Tracking Stability. In: Proceedings of 13th IWMS Vancouver, pp. 41–52 (1997)
18. Weaver, W., Timoshenko, S., Young, D.: Vibration Problems in Engineering. Wiley, New York (1990)
19. Orlowski, K., Wasielewski, R.: Washboarding during Cutting on Frame Sawing Machines. In: Proceedings of 15th IWMS Los Angeles, pp. 219–228 (2001)

Chapter 7
Tool Wear

Woodworking tools generally operate at high speeds (up to 50–80 m/s), their working surfaces (rake face, edge and clearance face) are subjected to large pressures due to cutting forces. The high speed motion generates high friction power turning into heat. Therefore, woodworking tools operate under heavy mechanical and thermal loads.

The cutting process is always associated with tool wear. Tool wear is actually a loss of material from the cutting edge due to mechanical, thermal and chemical effects. The wear process leads to bluntness of the cutting edge, which in turn causes an increase in power consumption, feed force and surface deterioration of the workpiece. Surface roughness and force components are especially sensitive to changes in edge profile geometry due to wear.

The appropriate tool in the woodworking industry achieve a high wood machining capacity, high dimension accuracy, good surface roughness and long service life. The costs allocated to buying and maintaining tools is significant in relation to the total production costs. Therefore, it is important to make the optimum selections of tools for different woodworking operations, and to use them rationally to ensure a long service life.

Tool wear is a very complex process due to the large number of constituents in wood and the presence of water and all of its details are not yet fully clear. Therefore, most of our knowledge concerning wear regularities and relationships are supplied by measurement results.

7.1 Edge Profile and Change in Cutting Force

The initial edge profile is generally produced by grinding. Depending on the properties of the material, (first of all its hardness,) this profile is never smooth and has an irregular geometry. The initial profile of the tool edge is not smooth enough, and the form (concerning its stress state) is not optimal. After a tool is put into use, its edge profile undergoes a rapid change and takes on the shape of a smooth

E. Csanády and E. Magoss, *Mechanics of Wood Machining*,
DOI: 10.1007/978-3-642-29955-1_7, © Springer-Verlag Berlin Heidelberg 2013

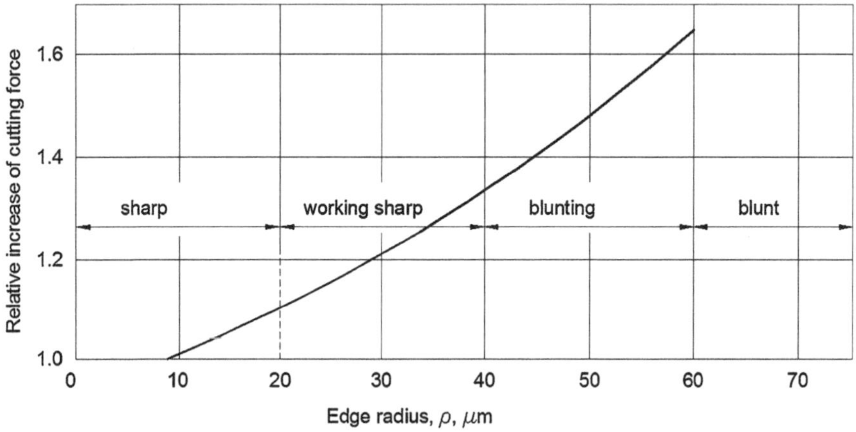

Fig. 7.1 Stages of tool wear and the relative increase of the required cutting force

cylinder. The wear processes cause the radius of the tool edge to continuously increase and several stages of the tool wear can be distinguished. The initial edge radius, depending on the hardness of the tool material, is generally to 10–15 μm. As the initial stage of tool wear developed, the optimum edge shape extends to an approximate radius of 20 μm (Fig. 7.1).

The next and more important stage is the "working sharp" stage, which ensures the optimum cutting conditions with low energy consumption and good surface quality. Over an edge radius of 40 μm, the tool starts to become blunt, requiring increasing cutting forces and decreasing the surface quality. The tool in this stage of wear is suitable for rough woodworking operations, but it does not ensure a good surface quality. Finally, in the blunt wear stage the tool requires high cutting energy and produces low surface quality. In addition, the blunt edge compresses the upper layers of the cut surface causing severe damage and surface instability.

Quantitative measurement of tool wear is of great importance to the woodworking industry. Wear can be defined as a loss of material from the cutting edge due to mechanical, electro-chemical and high temperature effects associated with the cutting process. The direct measurement of material losses would be complicated. Therefore, different parameters of the edge profile are used. The shape of the worn edge may be symmetric or asymmetric. In the simpler symmetric case, there is a uniquely defined relationship between the edge radius ρ and the initial position of an ideally sharp cutting edge y (Fig. 7.2):

$$\rho = y \frac{\sin\left(\frac{\beta}{2}\right)}{1 - \sin\left(\frac{\beta}{2}\right)} \tag{7.1}$$

where β is the sharpening angle of the tool edge.

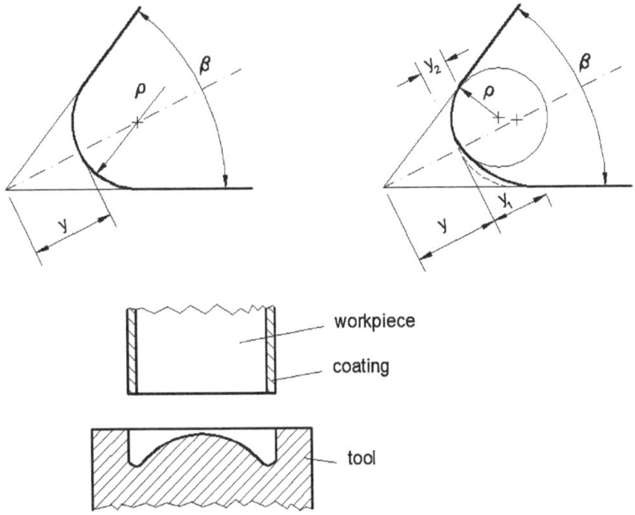

Fig. 7.2 Different edge wear profiles: **a** symmetric, **b** asymmetric, **c** wear due to workpiece coating

Due to the higher pressures acting on the bottom (clearance) face of the edge, the wear profile generally assumes an asymmetric shape (Fig. 7.2b) allowing other measurement parameters to be used. A cylindrical edge surface will also be developed here, but its centre point lies slightly over the bisector.

Particleboard and coated wood materials have an uneven density and hardness distribution; therefore, their wear action can be quite different along their cross section (Fig. 7.2c). Some glues may have strong abrasive properties.

The most commonly measured wear parameter is edge retraction y, which can easily be measured. This single wear parameter does not provide full information about the shape and geometry of a worn edge. It was also observed that the nose radius (Fig. 7.2b) may be stabilized at a particular value and does not always increase further with increasing wear.

It is also important to note that the characteristic wear profile shown in Fig. 7.2b remains similar using different tool materials, such as tool steel, high speed steel and carbide tools. That also applies to dry and green wood or particleboard.

The quantifying of tool wear when the wear varies along the cutting edge is more complicated. One possible method is to calculate the area under the wear profile along the edge. A simpler way is to consider the wear only below the outer surfaces of the board where the density is the highest and the maximum wear of the tool occurs. It is found that nose (or rake) recession has a good correlation with the area under the nose recession curve drawn along the cutting edge [1].

Tool wear can be related to the feed distance (L_f) and the true cutting length (L_c). The latter can be calculated as (see Fig. 7.3):

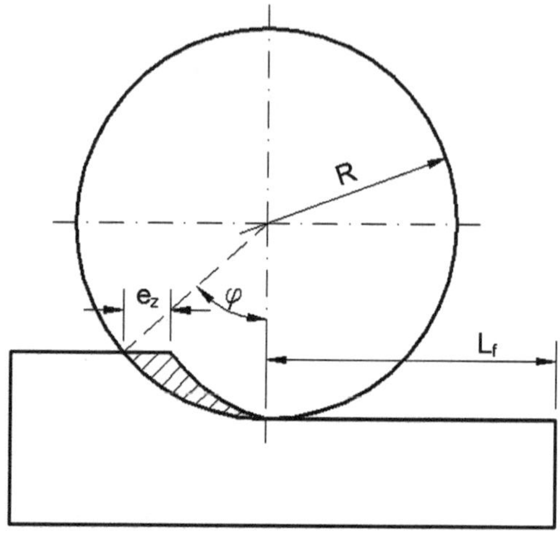

Fig. 7.3 Calculation of true cutting length

$$L_c = L_f \frac{R \cdot \phi}{e_z \cdot z} \tag{7.2}$$

where R is the tool radius, e_z is the tooth bite, z is the number of teeth and the angle φ must be substituted in radian.

The true cutting length when peeling wood will simply be the veneer length produced.

7.2 The Major Tool Wear Mechanisms

Three major wear mechanisms can be distinguished in tool wear: abrasion, electrochemical and high temperature corrosion. Generally, abrasion is always present as a mechanical wear process acting together with one of the other two, depending on the density, moisture content, and acidity of the wood and cutting parameters, especially edge surface temperature; (see in Sect. 2.3).

Abrasion occurs from the motion of hard particles between the sliding surfaces of a tool and wood. Wood species have always some mineral contaminations, which can be estimated by using a combustion method. The remaining hard mineral contaminations can be separated into fractions and measured. Their particle size is generally smaller than 50 μm. Silica is especially responsible for excessive blunting of tools. The silica content may be very different ranging from 0.1 to 10 g/kg wood.

The density of wood is also important in the process of mechanical wear. The denser wood has higher mechanical strength and exerts higher pressure on the edge face of a tool.

The amount of abrasive wear is proportional to the length of sliding contact between the tool and workpiece:

$$Q = A \cdot l$$

where A is a constant.

The rate of wear due to abrasion is given by differentiating the above equation

$$\frac{dQ}{dt} = A\frac{dl}{dt} = A \cdot v$$

This equation shows that the rate of wear is linearly related to the cutting speed.

Electro-chemical wear is attributed to the extractives in wood, such as gums, resins, sugars, oils, starches, alkaloids and tannin in the presence of water. Most extractives are reactive compounds forming chemical reactions between the wood extractives and the metallic constituents of a tool. The acidity of wood (pH-number) has great importance in this wear process. The cutting process is always associated with electric discharges contributing to the wear process. Early experiments have shown that negative electrical potentials imposed between the tool and workpiece can inhibit tool wear [2, 3], while positive electric potentials increased the wear rate.

The amount of electro-chemical wear depends on the length of time that the tool-electrolyte interaction occurs and it can be described by the following simple equation:

$$Q = B \cdot t$$

The wear rate is constant for a given combination of tool and workpiece:

$$\frac{dQ}{dt} = B$$

The constant B depends on wood species, moisture content and especially on its acidity [4, 5]. For example, oak and incense cedar (Calocedrus) have low *pH*-values.

The metal components in a tool also have a definite effect on the wear rate. High alloyed *HSS* steel is more susceptible to chemical attack from the extractives than WC and Co in tungsten carbide.

High temperature corrosion is also a common wear mechanism due to high temperatures in the cutting zone. High temperatures around 800–900 °C may occur depending on the cutting speed and other operational parameters. This high temperature can induce oxidation, which attacks the tool material. Different wood species have different behaviour in high temperature oxidation. The intensity of high temperature corrosion also depends on tool material and wood species. Using cemented tungsten carbide tools, it was found that the main corrosion mechanism is the oxidation of the cobalt binder which deteriorates the structure of the material in the surface layer. Hot corrosion is generally associated with the formation of salt deposits on the surface. A close examination of these deposits has revealed that wood constituents, such as *Ca* and *K* are present on the knife edge. Both impurities *Ca* and *K* are known to accelerate hot corrosion processes. Furthermore, the deposits always contain a considerable amount of carbon, which also accelerate hot corrosion.

Making electron micrographs of *HSS* tool edges showed that apparent smearing and plastic deformation are present indicating high temperatures and pressures which are contributing factors to mechanical wear [6].

Each of the wear processes described above may affect tool wear to a varying degree depending upon the specific machining conditions. They may interact among themselves or be part of other processes. For example, additional abrasion may result from oxides, carbides, nitrides and other compounds formed from high temperature corrosion processes.

Considering high temperature wear, one should also take into account the effect of adhesives, finishes and other materials used to improve properties of wood products. Chloride and sulphate salts found in adhesives accelerate hot corrosion. It is not surprising that the machining of particleboard containing adhesive materials is always associated with high wear.

7.3 Factors Affecting Wear

To reduce tool wear, the selection of appropriate tool materials or their treatment, such as coatings, is very important. Coatings are more resistant to high temperature corrosion and provide a longer tool life. Polycrystalline diamond is highly resistant to abrasion and to chemical attack. A high coefficient of heat conduction in a tool material is of great importance in lowering the surface temperature in the cutting zone. A larger sharpening angle also lowers the maximum edge temperature; (see in Sect. 2.3).

Edge reduction due to wear can be expressed by the following empirical equation:

$$y = y_0 + A \cdot v^n \cdot L_c^m \tag{7.3}$$

where y_0 is the initial edge reduction after grinding, v is the cutting speed. The constants A, n and m can be determined experimentally.

The total feed distance during the life time T of a tool is:

$$L_{fT} = e_z \cdot n \cdot z \cdot T$$

where e_z is the tooth bite, n is the rotation speed and z is the number of teeth, while the true cutting distance during its life time is:

$$L_{cT} = n \cdot R \cdot \varphi \cdot \pi = \frac{60}{2\pi} v \cdot \varphi \cdot T \tag{7.4}$$

where n is the rotation speed per min.

Using Eq. (7.3) and expressing L_C, we obtain

$$L_c = \left(\frac{y - y_0}{A}\right)^{\frac{1}{m}} \cdot \frac{1}{v^{\frac{n}{m}}}, \tag{7.5}$$

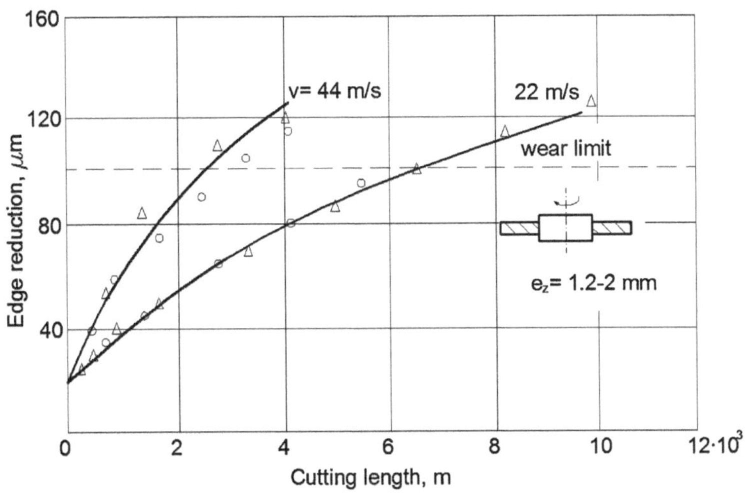

Fig. 7.4 Edge reduction as a function of cutting length for different cutting velocities [7, 8]

Taking a wear limit, of 100 μm, and combining Eqs. (7.4) and (7.5), the following general expression is valid (Taylor tool life equation):

$$T \cdot v^{\frac{n}{m}+1} = constant \tag{7.6}$$

or

$$v \cdot T^{\frac{1}{\frac{n}{m}+1}} = constant \tag{7.7}$$

Figure 7.4 shows experimental results for particleboard using a tungsten carbide tool ($\beta = 55°$, 1.2 mm tooth bite, 2–3 mm cutting depth, 60 mm tool radius) [7, 8].

The initial edge recession is $y_0 = 20$ μm ($\rho = 17$ μm) and the calculated exponents in Eq. (7.3) are $n = 0.74$ and $m = 0.65$. Using these exponent values, the Taylor equation assumes the form

$$T \cdot v^{2,14} = const$$

or

$$v \cdot T^{0,47} = const$$

Using logarithmic co-ordinates, the Taylor equation is represented by a straight line (Fig. 7.5), which can easily be used to determine the service life after resharpening a tool.

The maximum cutting length during the service life of a tool is (see Eq. (7.5)):

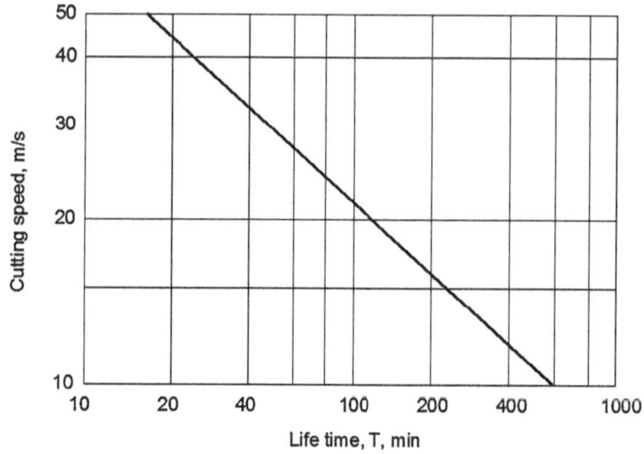

Fig. 7.5 Taylor life time curve (Based on data in Fig. 7.4)

$$L_{cT} = \frac{const}{V^{\frac{n}{m}}}$$

The exponent *n/m* = 1.14 in Fig. 7.4 shows that with increasing cutting speed, the maximum cutting length decreases. The exponent in the Taylor tool life equation has different values depending on the dominant wear process (abrasive, electro-chemical or hot corrosion) in a particular case. In pure abrasion, the exponent of *T* is equal to one; for electro-chemical wear, this exponent is greater than unity and for high temperature wear, it is less than unity. The maximum cutting length during the service life of a tool may be independent or dependent on the cutting speed.

Figure 7.6 shows a Taylor tool life diagram for Ponderosa pine in the low cutting speed range [4].

The exponent of *T* is greater than unity depending on the moisture content. That means that the dominant wear mechanism is the electro-chemical wear. The low cutting speed ensures moderate heat generation on the tool edge avoiding high temperature corrosion. Cutting air-dry Ponderosa pine, however, the value of exponent is 1.1 quite near to the unity. Therefore, abrasion seems to be the dominant wear mechanism in this case. As a consequence, the maximum cutting length during the service life is more or less independent of cutting speed.

The most common tool materials used are the low alloy steel, *HSS* steel, stellite tipped tool, cemented tungsten carbide, boron nitride and polycrystalline diamond. Various edge treatment and coating procedures were examined for further increase of the service life of tools [9]. The above list of tool materials means, at the same time, the sequence of increasing hardness and tool service life. Hard materials are generally more rigid and breakable. Hard materials require a bigger sharpening angle and, as a consequence, a smaller rake angle. It is important to note that the initial sharpness (y_0) of a tooth made of hard material is always inferior to edges made of low alloy or *HSS* steel. Peeling knives require small sharpening angle around 20° and a good initial sharpness. A special coating of the steel knife will maintain satisfactory sharpness of a tool with a long service life [10].

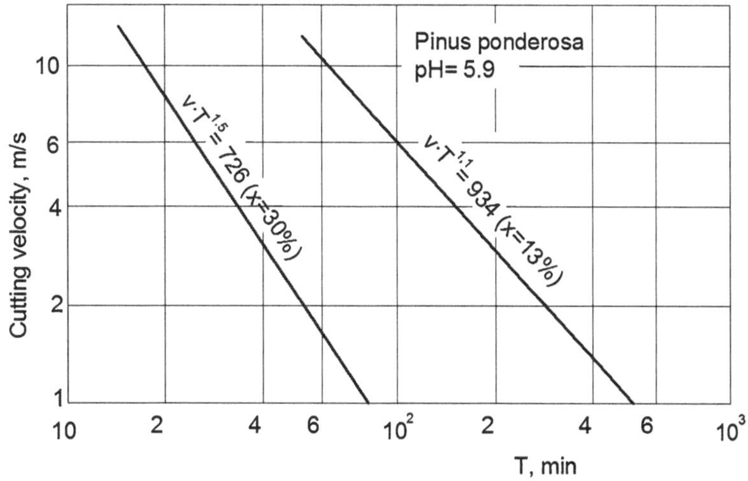

Fig. 7.6 Tool life diagram for wet and air-dry Ponderosa pine in the low cutting speed range [4]

Finally, the main factors influencing tool wear can be summarized as follows:

- **Tool properties**
  - tool material
  - sharpening angle
  - rake angle
  - edge treating or coating
  - accuracy of edge running circle
  - accuracy of side running
- **Workpiece properties**
  - strength properties
  - moisture content and *pH*-value
  - density
  - cutting direction
  - resin and ash content
  - adhesives and surface coating
- **Operational parameters**
  - cutting speed
  - tooth bite
  - cutting depth
  - vibration of the tool

The number of influencing factors is very high, the interaction of different wear mechanisms is very complicated and, therefore, the tool life for a particular case can only be determined experimentally. In practice it can often be observed that more frequent re-sharpening of the tools is carried out to secure good surface quality under any circumstances. It is clear that the combine requirement of quality and economy can be achieved by compromise in the wise choice of influencing factors.

References

1. Sheikh-Ahmad, J., McKenzie, W.: Measurement of tool wear in the machining of particleboard. In: Proceedings of 13th IWMS Vancouver, pp. 659–670 (1997)
2. Kivimaa, E.: Was ist die Abstumpfung der Holzbearbeitungswerkzeuge? Holz als Roh- und Werkstoff, S. 425–428 (1952)
3. Алексеев, А., Влияние электрических явлений на износ инструментов. (Effects of electric phenomena on tool wear) Дерев. Пром., (8), с. 15–16 (1957)
4. Tsai, G., Klamecki, B.: Separation of abrasive and electrochemical tool wear mechanisms in wood cutting. Wood Sci **4**, 236–241 (1980)
5. Packman, D.: The acidity of wood. Holzforschung, 14. S. 179–183 (1960)
6. Stewart, H.: Feasible high-temperature phenomena in tool wear. Forest Product J. **39**(3), 25–28 (1989)
7. Saljé, E., Dubenkropp, D.: Das Kantenfräsen von Holzwerkstoffplatten. Holz- und Kunstoffverarbeitung. No. 4. (1983)
8. Saljé, E., Drückhammer, J.: Qualitätskontrolle bei der Kantenbearbeitung. Holz als Roh- und Werkstoff, S. 187–192 (1984)
9. Gottlob, W., Ruffino, R.: AISI D-6 steel and Stellite-1 tool life determination. In: Proceedings of 13th IWMS Vancouver, pp. 761–769 (1997)
10. Beer, P.: Synergism of the abrasive and electro-chemical wear of steel tools in wood peeling. In: Proceedings of 15th IWMS, Los Angeles, pp. 103–108 (2001)

Chapter 8
Surface Roughness

Machined surfaces—regardless if they are made of metal, plastic or wood—are never perfectly smooth with protruding parts, valleys and peaks. These forms of surface irregularity are called *roughness*. Surface roughness can be caused by different factors: discontinuities in the material, various forms of brittle fracture, cavities in the texture (e.g., wood), radius of the tool edge, local deformations deriving from the free cutting mechanism (without counter blade).

Surface roughness usually has a primary influence on the visual appearance of materials but it might have other effects, too. Surface roughness may be extraordinarily detrimental for wood if the surface under the tool edge suffers permanent deformity. The stability of the damaged surface diminishes to a great extent; the durability of the processed surface will then be inferior.

Visual appearance and colour effects are primarily influenced by dispersion and reflection of light. An apparent example for this is transparent glass, which loses transparency following a moderate roughening process. The original colour of wood becomes a lot more visible if the surface is "bright," (smooth and free of irregularities). Speaking of wood, a good example is ebony. The black colour gives entirely different effects depending on the surface roughness. The polished surface presents a bright black colour. A surface treated with colourless lacquer, oil or wax will lead to quite different colour effects again.

The minimum surface roughness that can be achieved depends on a number of factors. Generally we can say that processing materials with bigger volume weight can result in smoother surfaces. This explains the excellent polishing ability of ebony.

The surface roughness of wood results from many influencing factors, therefore defining general rules has taken quite a long time [1–3]. Several researchers, trying to overcome the difficulties, used a special filtering method to remove the deep valleys (vessels) from the surface profiles to increase the sensitivity of the roughness parameters to changes in the operational parameters [4–6]. Nevertheless, different wood species hardly could have been compared to one another concerning their surface roughness.

E. Csanády and E. Magoss, *Mechanics of Wood Machining*,
DOI: 10.1007/978-3-642-29955-1_8, © Springer-Verlag Berlin Heidelberg 2013

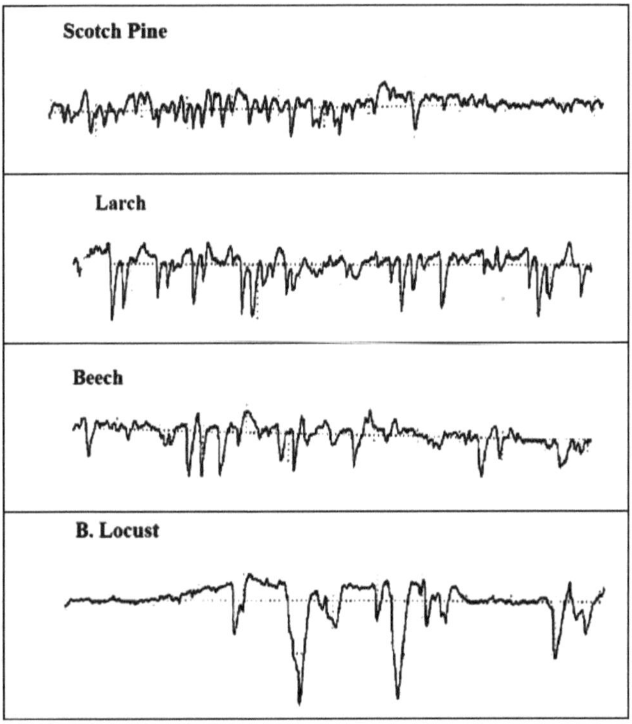

Fig. 8.1 Surface roughness profiles [8, 12]

New ideas in the latest decade and modern measurement techniques supported the research works to establish the main governing rules describing the behaviour of different wood species concerning their surface roughness in a unified system [7–12]. In the earlier time it encountered difficulties to characterize wood species and to express their structural properties in terms of numerical values. In order to overcome this problem, it was necessary to elaborate and introduce a new wood structure characterisation method. The newly defined anatomical structure number made it possible to treat all wood species in a unified system facilitating the recognition of general regularities.

8.1 Parameters of Surface Roughness

Machined surfaces always show irregularities of different height and depth that we call "roughness".

Typical surface roughness profiles for softwood and hardwood species are shown in Fig. 8.1. The first curve demonstrates the surface roughness of Scotch pine of thick growth, where both high and deep irregularities are small. The fourth

R_z	Height of profile elements DIN EN ISO 4287	$R_z = \frac{1}{5} \cdot (Z_1 + Z_2 + Z_3 + Z_4 + Z_5)$ 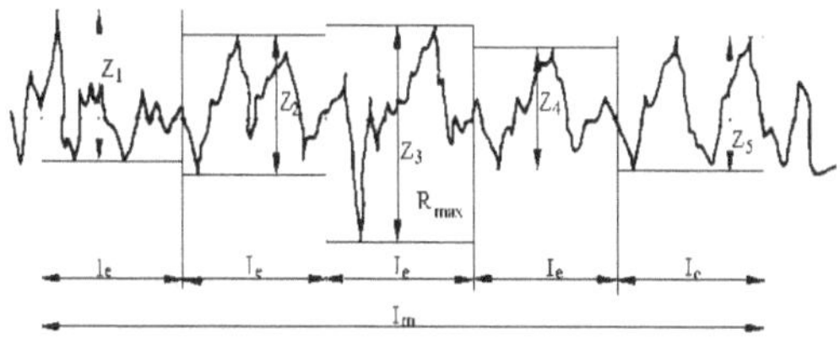
R_{max}	Maximum roughness profile depth DIN EN ISO 4287	
R_a	Average roughness DIN EN ISO 4287	$R_a = \frac{1}{l_m} \int_0^{l_m} \lvert z(x) \rvert \cdot dx$ 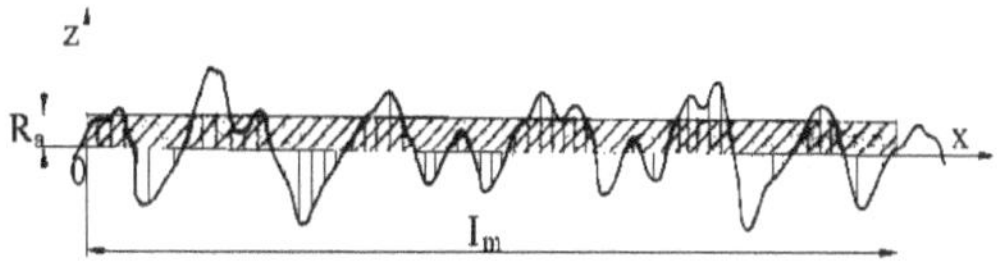
R_{zi} R_p R_v	Profile depth DIN EN ISO 4287 Roughness profile peak height DIN EN ISO 4287 Roughness profile valley depth DIN EN ISO 4287	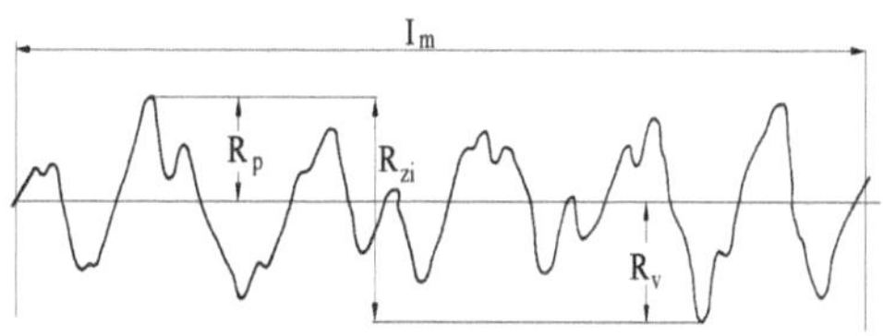
W_t	Wave depth DIN EN ISO 4287	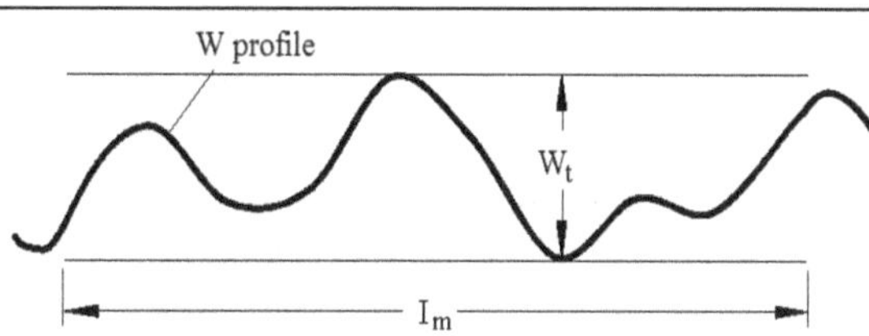

Fig. 8.2 Standardized parameters of surface roughness and waviness

curve illustrates the machined surface of black locust with large vessels, where the height of irregularities are below 10 μm but the large vessels cause depth irregularities between 50–80 μm.

It is impossible to find a roughness parameter, which gives an unambiguous characterization of the surface from all aspects; therefore many parameters have

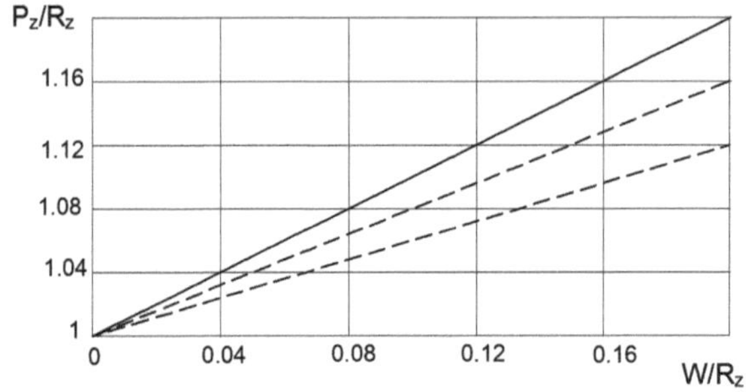

Fig. 8.3 Roughness of the unfiltered profile in relation to waviness based on Eq. (8.1), continuous line = theoretical correlation; area between dashed lines = range of the measured values [4]

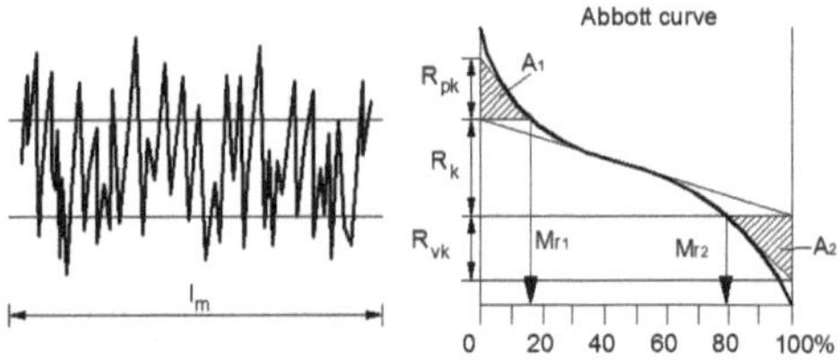

R_k –depth of the so-called 'core'
R_{pk} – reduced peak heights
R_{vk} – reduced valley depths
A_1 – material ratio of peak elements
A_2 – surface of valleys
M_{r1} – 'smaller' material portion value of the core
M_{r2} – 'higher' material portion value of the core

Fig. 8.4 Calculation and characterization of the material ratio curve (Abbott curve) DIN 4776; DIN EN ISO 4287

been derived, which have been standardized for the purpose of consistent interpretation and usage (Fig. 8.2).

The truly unfiltered profile also contains accidental waviness on the surface. The surface roughness depth P_z can be applied also for this profile, which has a theoretical correlation to the R_z value of the filtered profile as follows:

$$P_Z = R_Z + W$$

or

$$\frac{P_Z}{R_Z} = 1 + \frac{W}{R_Z} \qquad (8.1)$$

In practice maximum roughness profile heights do not necessarily fall on peak heights or valley depths, so the angle of inclination of the straight line described by Eq. (8.1) will be smaller; that is:

$$P_Z/R_Z = 1 + \alpha W/R_Z$$

where the value α is less than 1 (generally 0.6–0.8). The range of the measurement values is shown in Fig. 8.3 [4].

A part of the space within the roughness profile is filled with material; the rest of the space is filled with air. The relationship between these two factors is expressed by the material ratio curve of the profile, as in distribution curves. The material ratio curve of the profile is also known as the Abbott curve; its definition and parameters are shown in Fig. 8.4. Modern instruments perform this assessment automatically; the data are drawn and can be printed.

8.2 The Internal Structure of Wood

One of the typical characteristic features of wood is its internal structure, which has cavities in the form of vessels and cell lumens inside. The typical internal structure of softwood and hardwood species is shown in Fig. 8.5.

The early wood tracheids of Scotch pine have large cavities (20–40 μm) and thin walls; however cavities in the late wood are roughly half of the size. The structure of hardwood species is more complicated, consisting of a number of different cell types. Vessels consisting of vertical units doing the transportation; the diameter of which can be up to 300 μm; thus they are visible to the eye. The tracheids, cavities in the long parenchyma cells, are relatively small, in the range 10–20 μm; (see Table 8.1).

The cavities of both in the vessels and fibres in the early and late wood are also different in size. Furthermore—depending on the weather circumstances—this variability is typical for the internal structure of subsequent annual rings.

During mechanical processing of wood holes are cut in different angles, therefore even in the case of damage-free cutting (sharp cutting line) hollows do remain on the surface. These valleys cause a certain roughness on the surface, which is not effected by the machining process. Therefore the roughness evolving this way is called structural or structure-caused roughness.

From the aspect of roughness, the internal structure of wood is characterized by the mean diameter of holes and the number of holes in the particular cross-section unit. The size and number of holes has to be determined both in the early and late wood, therefore, the early and late wood ratio must also be established.

In order to gain the above data, small-sized specimens are taken from different wood species, where the required data are established using a measuring microscope or by means of digital image processing. It is practical to check the obtained data also by calculation methods. The bulk density of the sample is easy to

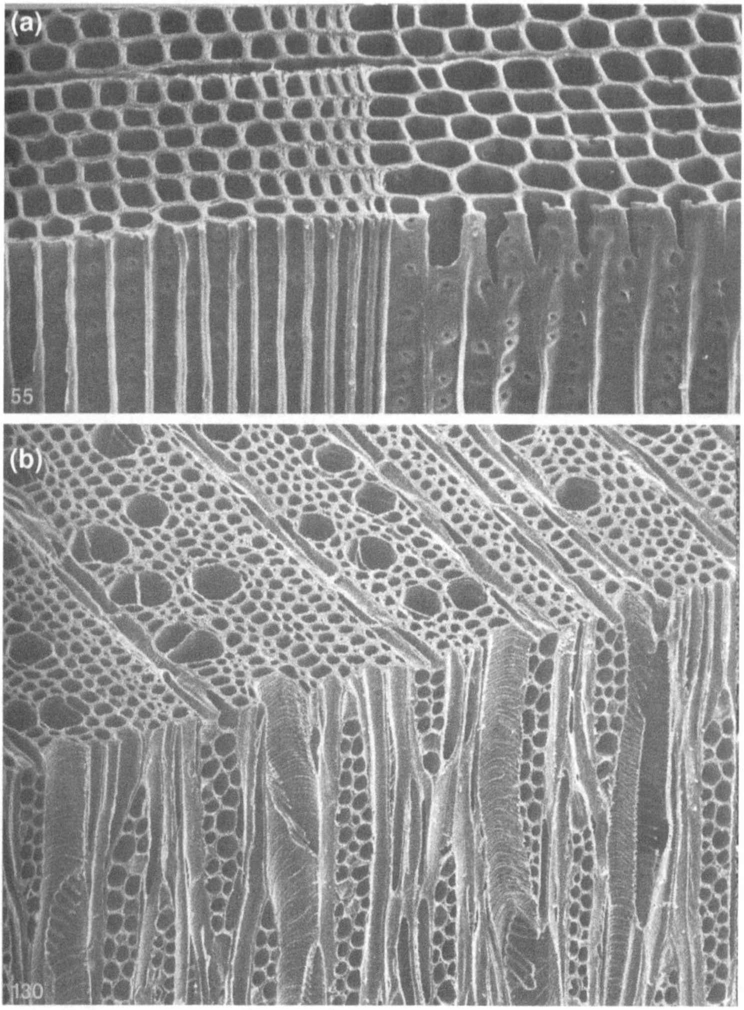

Fig. 8.5 Microscopic photo of pine (**a**) and hardwood (**b**) species [13]

determine; the following approximation equation has to be effective (based on a 1 cm^3):

$$\rho_t = \left[\left(1 - \frac{d_1^2 \cdot \pi}{4} \cdot n_1\right) \cdot a + \left(1 - \frac{d_1^2 \cdot \pi}{4} \cdot n_1\right) \cdot b\right] \cdot \rho \qquad (8.2)$$

where

ρ_t, ρ —bulk density and real density of wood,
d_1, d_2 —mean diameter of cavities in the early and late wood
n_1, n_2 —number of cavities on the unit surface in the early and late wood
a, b—ratio of early and late wood ($a + b = 1$).

Table 8.1 Structural properties of specimens

Wood species	Early wood			Late wood		
	$\overline{d_i}$ [μm]	$\overline{n_i}$ [piece/cm^2]	$\overline{a}$	$\overline{d_i}$ [μm]	$\overline{n_i}$ [piece/cm^2]	$\overline{b}$
Thuja	26.5	142 800	0.8482	14.0	316 600	0.1518
Spruce	30.0	111 335	0.8478	19.0	160 400	0.1522
Pine	28.0	125 100	0.6694	20.0	135 840	0.3306
Larch	38.0	65 490	0.6310	17.5	145 000	0.3690
Beech (vessel)	66.0	15 740	0.7000	48.0	14 020	0.3000
Beech (tracheid)	8.2	342 890		6.4	490 290	
Oak (vessel)	260.0	400	0.5900	35.7	12 000	0.4100
Oak (tracheid)	22.5	120 000		19.6	85 000	
B. locust (vessel)	230.0	546	0.5800	120.4	1 500	0.4200
B. locust (tracheid)	15.0	270 000		9.6	280 000	
Cottonwood (vessel)	69.7	9 500	0.6666	44.0	12 700	0.3333
Cottonwood (tracheid)	12.7	309 500		11.0	300 892	
Ash (vessel)	177.0	670	0.6100	52.00	750	0.3900
Ash (tracheid)	19.0	190 000		14.0	230 000	

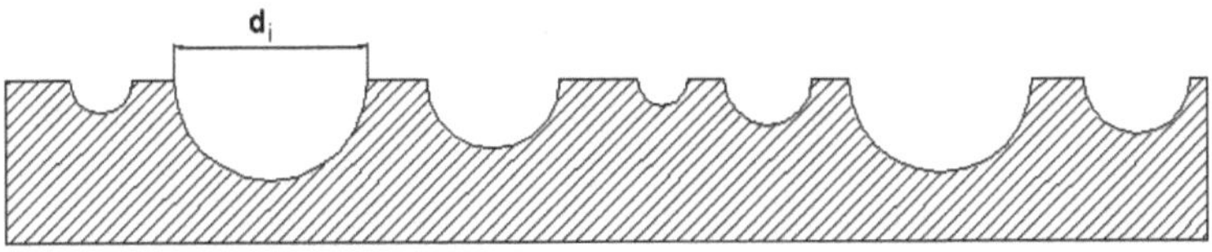

Fig. 8.6 The model of structural surface roughness [7]

We can use the above equation to double-check the early and the late wood separately, knowing that the bulk density of late wood is approximately two times higher than that of early wood.

Table 8.1 summarizes the typical characteristics of the internal structure of samples of conifers and hardwoods [7].

Cutting vessels, tracheids and other elements of the texture causes surface irregularities. We quantified the number of vessels cut on a certain length in the direction of processing. The scattering of the vessel diameters usually shows normal distribution, which enables the utilization of the mean diameter size without making a greater error.

The position of the vessels measured on the surface is always accidental, which obviously causes a scattering of the surface roughness.

Adding up the number of structural elements cut on the surface gives a measure of surface roughness as shown in the model in Fig. 8.6.

The area of the valleys has a connection with the number and diameter of structure elements measured along a given unit in the processing direction, which is expressed in the following equation:

$$\Delta F = \frac{\pi}{8}\left[a.\left(\sqrt{n_1}\cdot d_1^2 + \sqrt{n_1}\cdot d_2^2\right) + b\cdot\left(\sqrt{n_3}\cdot d_3^2 + \sqrt{n_4}\cdot d_4^2\right)\right][\mathrm{cm}^2/\mathrm{cm}] \quad (8.3)$$

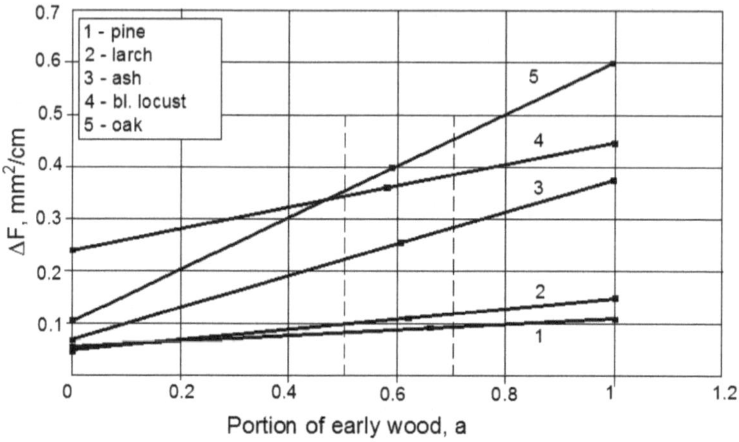

Fig. 8.7 Alteration of the ΔF value of different wood species in relation to the early wood ratio [11]

where

n_1, n_2 —the number of vessels and tracheids in the early wood in the unit cross section
n_3, n_4 —the number of vessels and tracheids in the late wood in the unit cross section
d_1–d_4— the diameter of vessels and tracheids in the early and late wood,
a, *b*—ratio of early and late wood

The value ΔF defined with the Eq. (8.3) is called *structure number*, which gives an accurate definition of each wood species based on the size and number of cavities in the wood structure. Surface irregularities caused by the internal wood structure are expected to have a definite correlation with the structure number.

A further advantage of the structure number is that it enables the characterisation of wood species based on their internal structure, and it helps to establish correlations of among the surface roughness parameters.

It is well-known that results of surface roughness tests usually show significant scatterings. One reason is the accidental position of tracheids and vessels on the machined surface. A more significant scattering can be caused by the random position and cut of the early and late wood or the seasonal change of the early wood/late wood ratio. The value ΔF may have substantial differences in the early and late wood. Pine species show the smallest divergence, whereas hardwood species generate a significantly bigger one. Figure 8.7 illustrates the alteration of the ΔF value of different wood species depending on the early wood ratio. The starting point of the curves ($a = 0$) indicates pure late wood, while the end point ($a = 1$) indicates pure early wood. Oak shows the biggest change, but ash shows a significant change as well.

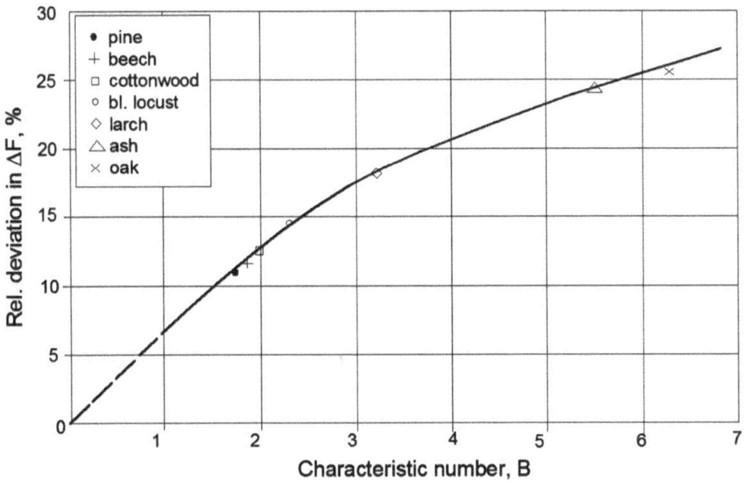

Fig. 8.8 Relative changes of ΔF in relation to the parameter B [11]

Tests show that the early wood ratio predominantly ranges from $a = 0.5$–0.7 as shown in Fig. 8.7.

Therefore it seemed practical to apply the relative changes of ΔF in the range a = 0.5–0.7 with the following equation:

$$\delta(\Delta F) = \frac{\Delta F_{0.7} - \Delta F_{0.5}}{\Delta F_{0.6}} \tag{8.4}$$

Wood species can be characterized with the ratio of cross sections of cavities cut in the early and late wood, which we can express similar to Eq. (8.5) as follows:

$$B = \frac{d_1^2 \cdot \sqrt{n_1} + d_2^2 \cdot \sqrt{n_2}}{d_3^2 \cdot \sqrt{n_3} + d_4^2 \cdot \sqrt{n_4}} \tag{8.5}$$

Since pine species lack these vessels, Eq. (8.5) for conifers will be simpler.

Figure 8.8 shows the relative changes of ΔF in relation to the parameter B defined with Eq. (8.5).

The apparent correlation between the two variables is clearly displayed. However it is remarkable that beech is located right next to Scotch pine; and black locust shows a smaller relative change than larch.

Consequently the relative change has no connection with the absolute value of ΔF. The correlation follows the following empiric equation:

$$\delta(\Delta F) = 7.8 \cdot B^{0.65} [\%] \tag{8.6}$$

8.3 The Origin of Surface Roughness

Roughness that evolves during machining has two major components: machining-caused roughness and roughness caused by the internal wood structure. Even in the case of an ideal machining rough surface evolves due to the inner holes cut. Moreover, in the recent practice of high-speed machining, roughness due to machining is usually much less than the structure-based roughness, especially in the case of hardwood species with large vessels.

The roughness due to machining usually depends on the following factors: cutting speed, chip thickness, machining direction relatively to the grain, rake angle of the tool, sharpness of the tool edge (tool edge radius) and vibration amplitude of the work piece.

Wood cutting belongs to the group of the so-called free cutting. Its main characteristic feature is the absence of a counter-edge, therefore, the counterforce is produced by the strength of wood and forces of inertia. The higher strength and hardness parameters (modulus E) the wood has the smaller force of inertia is required; that is to say, the slower the roughness increases with the decrease of the cutting speed.

The primary reason for machining-caused roughness is the brittle fracture of the material and its low tensile strength perpendicular to the grain. The occurrence of brittle fracture cannot be eliminated, however, it can be limited to a lower volume. The most effective method for this is the high-speed cutting and the smallest material contact possible (sharp tool edge).

The only way to eliminate the negative effect of the tensile strength perpendicular to the grain is to generate a compression load in the immediate vicinity of the edge. This can be facilitated with a 65–70° tool angle or a 20–25° rake angle. Especially the edge machining of boards is very inclined to breaking the edge because of the tensile load; therefore the selection of appropriate kinematic conditions is very important (see Fig. 3.12).

Excessive compression load deforming the material can also cause roughness. The method "smooth machining" has been known for a long time, which is based on the knowledge that smaller chip thickness raises smaller forces. Compression load can also be reduced by using the "slide cutting". Slide cutting produces shear stresses on the edge, which also contributes to material failure. In accordance with the well-known equation the equivalent stress has the following form (see Sect. 2.6):

$$\sigma_e = \sqrt{\sigma^2 + 4\tau^2}.$$

The distribution of compression load under the tool edge depends on its radius (bluntness of the tool edge). Due to the finite thickness of the edge a given layer of thickness z_0 will be compressed underneath the edge (Fig. 8.9).

The thickness of this layer is about 70 % of the tool edge radius. The load at the edge is transferred onto the material on a surface that is *2b* wide and a length that is identical with that of the edge. This is similar to strip load. The biggest load

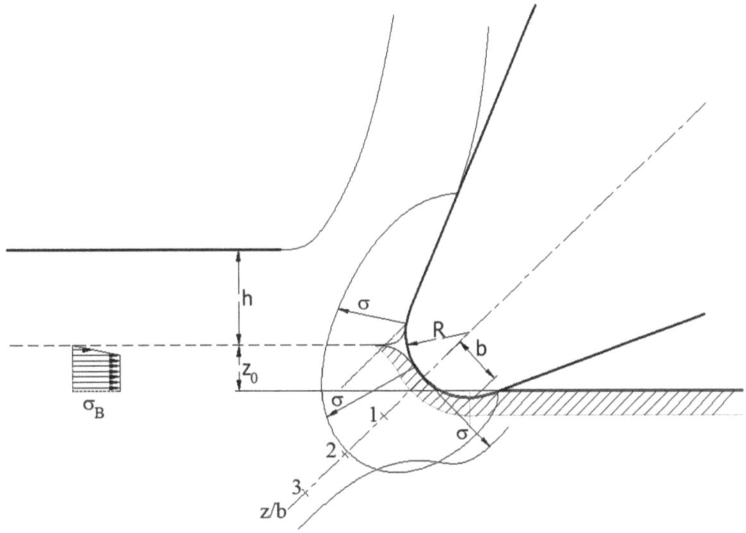

Fig. 8.9 Compression effect at the edge

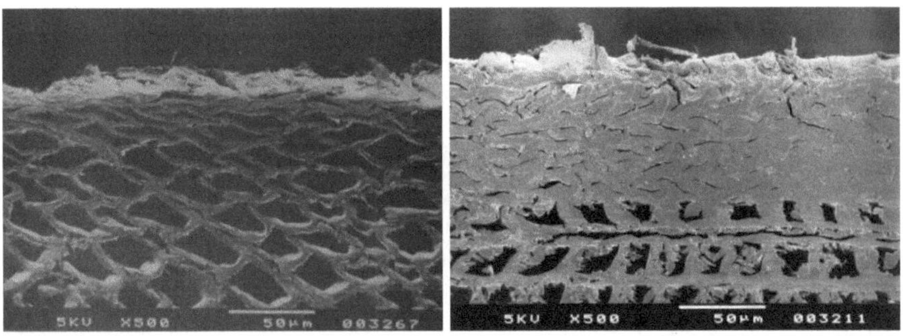

Fig. 8.10 Compaction behaviour of early and late wood layers caused by the same blunt edge [14]

appears just under the contact surface and it rapidly decreases as we move towards the inside of the material (load of the infinite half space). This means that the highest compression of cells always starts directly underneath the edge.

If the compression load underneath the edge reaches the ultimate bearing stress of the material, the cell system suffers permanent deformation and it gets compressed to the detriment of the cavities [14]. If out of total deformation z_0 the permanent deformation is z_1, then the expelled material will be located in the lower layers and the thickness of the compressed layer z_2 is (see Figs. 8.10 and 8.26):

$$z_2 = z_1 \cdot \rho_1 / (\rho_2 - \rho_1) \qquad (8.7)$$

where

ρ_1—is the volume density without compression
ρ_2—is the volume density of the compressed material (1–1.2 g/cm^3)

The volume density of early and late wood is significantly different, while the value ρ_2 can change only to a limited extent; therefore we can expect the following values after compression: approximately $z_2 = z_1$ in the early wood and $z_2 \cong 3z_1$ in the late wood.

This was also observed experimentally [14]. Figure 8.10 shows the compaction of early and late wood using the same blunt tool. The late wood was compacted several times deeper than the early wood.

If for the purpose of simplicity we presume that the elastic deformation z_r is sustained till the crushing stress σ_B is reached, then:

$$z_r = \frac{\sigma_B 4b(1 - \nu^2)}{E} \tag{8.8}$$

The deformation underneath the edge is elastic in the first period of the compression so it will rebound once the edge has passed. The approximate rate of elastic deformation can be calculated easily. When the elastic half space is exposed to a strip load:

$$\sigma = \frac{E}{2 \cdot (1 - \nu^2)} \cdot \frac{z}{2b} \tag{8.9}$$

In pine wood, the compression load between directions **B** and **C** can have values of $\sigma_B = 15$ N/mm^2; then a $R = 10$ μm radius is likely to produce elastic deformation of approximately $z_r = 1$ μm, while a $R = 50$ μm radius is expected to bring elastic deformation of approximately $z_r = 5$ μm. These values give one-seventh of the expected total deformation ($z_0 = 7$ and 35 μm,). The compression of the upper layer should be considered in each case.

The wood structure however contains cavities and even cavities of smaller size are comparable to the radius of a sharp tool edge. Therefore the sharp tool edge can penetrate the cavities and break the cell walls. The tool edge bends these broken wood parts, standing out from the cutting plane, which then take the necessary deformation z_0 without transferring it towards the lower layers (see Fig. 8.26a).

When the tool edge radius is increased, both the edge size and the rate of deformation z_0 exceed the size of cavities; consequently a compression of the surface layer develops. An approximately $R = 20$ μm tool edge radius is expected to trigger a surface compression.

In some respect the jointing of an edge has similar effects as edge wear. Jointing is the common practice to produce the same cutting circle for all knives of a cutter head. The jointed land at the edge has a zero degree clearance angle which will compress a thin layer of the wood causing cell damage depending on the jointed land width. This cell damage may be responsible for surface instability, a decrease

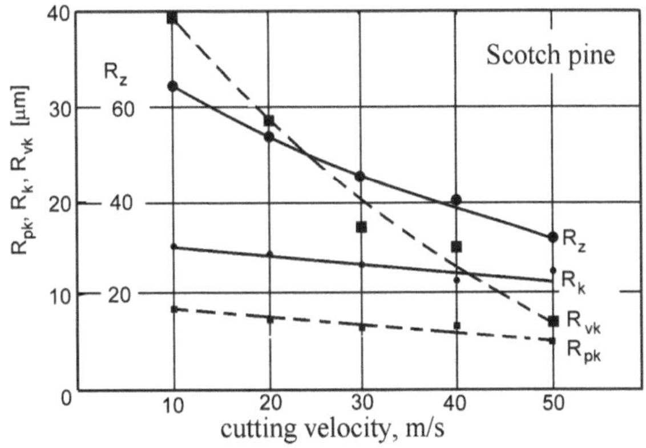

Fig. 8.11 Effects of cutting speed on the surface roughness parameters of Scotch pine [7]

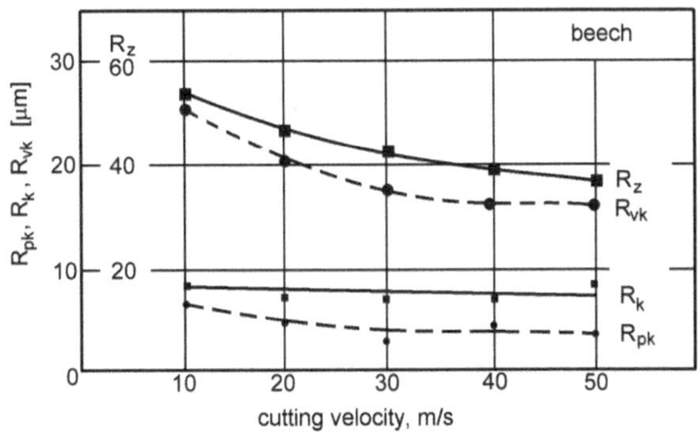

Fig. 8.12 Effects of cutting speed on the surface roughness parameters of beech [7]

in gluing strength. It is generally accepted that a narrow band width around 0.25 mm will not create many problems because the area of zero clearance is very small. It is also recommended that the jointed land width should not be more than 1.0 mm.

8.4 The Effects of Machining Process on Surface Roughness

It is well known that a higher cutting speed results in a smoother surface that means smaller roughness values in terms of both the average surface roughness R_a, and depth of irregularities R_z.

Expanding the assessment on the characteristics of the *Abbott curve* we receive more details concerning the components of the surface roughness (Figs. 8.11 and 8.12). The average diameter of vessels of beech was 60 μm, while for tracheids the corresponding value was 10–15 μm. The mean inner diameter of the tracheids of Scotch pine was 25–30 μm in the early wood and 13–18 μm in the late wood.

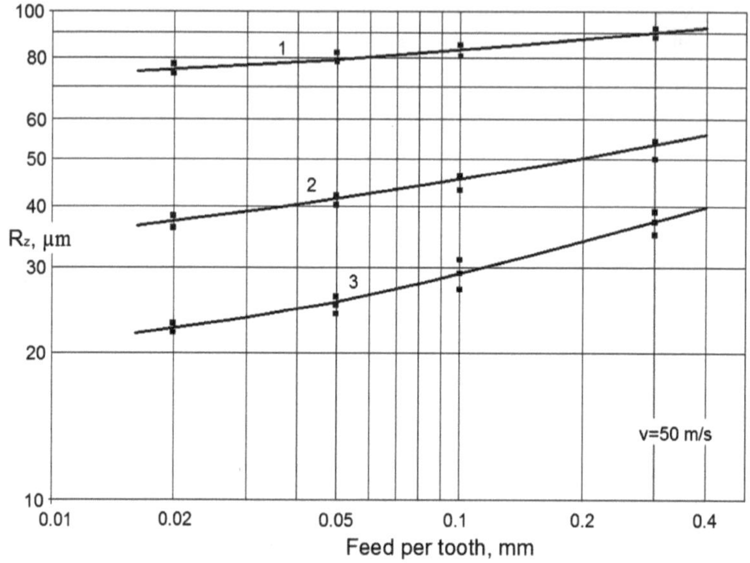

Fig. 8.13 Effects of feed per tooth (e_z) on irregularity depth (R_z) 1-oak; 2-beech; 3-Scotch pine [8]

From Figs. 8.11 and 8.12 in both cases the R_{pk} and R_k values remain nearly constant or slightly decrease as a function of cutting speed, while R_{vk}-values depend on cutting speed.

This dependence is stronger with pine, especially at low cutting speeds. This result may be explained by the fact that the pine wood had less local stiffness around the cutting edge, so inertia forces play a more important role to ensure a clear cutting surface. Beech has larger structural cavities, giving greater R_{vk}-values even at high cutting speeds.

Chip thickness or feed per tooth influences the surface roughness to a smaller extent. This is explained by the fact that the increase of the chip thickness also raises increased forces; and the thicker chip can transmit bigger forces in the connected area at the point where chips are detached.

Figure 8.13 shows the effect of feed per tooth (e_z) on the irregularity depth (R_z).

The softer the processed wood is, the bigger is the effect of feed per tooth. The combined parameter of the Abbott curve ($R_k + R_{vk}$) also shows a good correlation with the feed per tooth, as illustrated in Fig. 8.14. Increasing the feed per tooth also increases the reduced hollow depth ratio R_{vk}.

The curves in Fig. 8.13 can also be expressed as follows:

$$R_Z = A + B \cdot e_z^n$$

where A, B and n are constant values. The value of exponent is 0.6 for all of the three curves, and the value of B is almost identical.

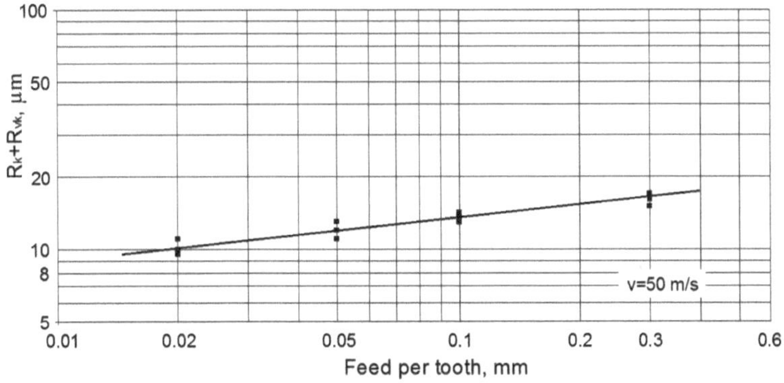

Fig. 8.14 Effects of feed per tooth e_z on the combined parameter $(R_k + R_{vk})$ for Scotch pine v = 50 m/s [8]

8.5 Internal Relationships Between Roughness Parameters

Examining the correlations between the common roughness parameters (average roughness R_a, irregularity depth R_z) and the Abbott curve we discover some interesting interrelations (Figs. 8.15 and 8.16). Figure 8.15 shows a strong relationship between the average roughness and the sum of Abbott parameters.

It is well-known that between R_a and R_z there is only a poor interrelation. As a consequence, no uniquely defined relationship between R_z and the sum of Abbott parameters can be expected. Nevertheless, the experimental results depicted in Fig. 8.16 show an interesting picture.

A lot of curves are obtained. For a more accurate explanation, the measurement results on MDF-boards of different volume density included [8]. MDF has the most uniform internal structure which gives the lowest curve. Oak has large vessels and a much less uniform structure and gives the uppermost curve. The curves for other species are between these two extremes according to their inhomogeneities. Most curves obey the following general form

$$R_z = A.\left(R_{pk} + R_k + R_{vk}\right)^{0.65} \tag{8.11}$$

and the constant can be expressed as

$$A = 7.45 \cdot (R_k + R_{vk})/R_z \tag{8.12}$$

Using the Abbott curve, we can determine the lack of material in the uneven surface. An equivalent layer thickness may be calculated (see Fig. 8.4) as follows

$$\Delta h_e = R_{pk} \cdot \left(1 - \frac{M_{r1}}{2}\right) + \frac{R_k}{2} + \frac{R_{vk} \cdot (1 - M_{r2})}{2} \tag{8.13}$$

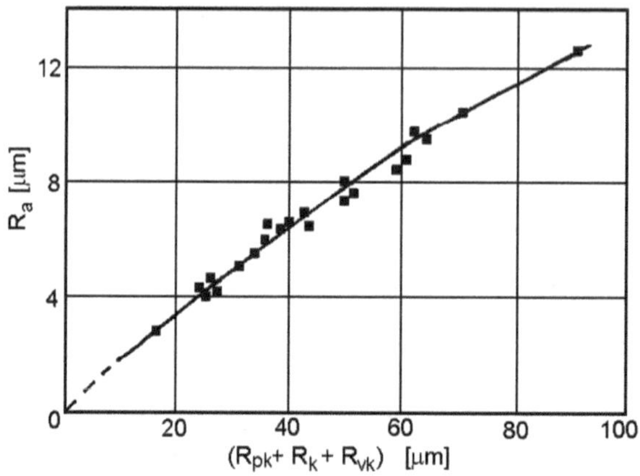

Fig. 8.15 Relationship between R_a and the Abbott-parameters [8]

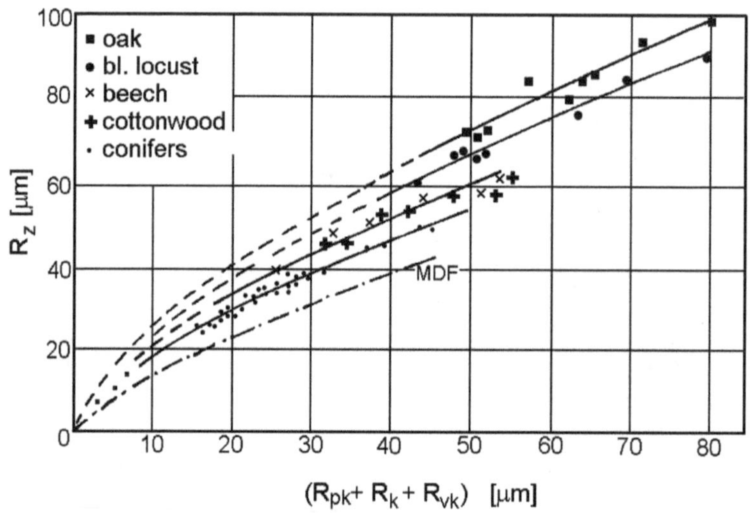

Fig. 8.16 Relationship between irregularity depth R_z and the Abbott-parameters [8]

where M_{r1} and M_{r2} should be substituted as decimal values. The following rough estimation shows the weight of the parts in Eq. (8.13):

$$\Delta h_e = 0.95 \cdot R_{pk} + 0.5 \cdot R_k + 0.08 \cdot R_{vK} \tag{8.14}$$

In practical cases the R_{pk}-layer can eventually be neglected because the few peaks sticking out from the surface can easily be crushed by pressing.

A graphic representation of the lack of materials related to the unit surface is seen in Fig. 8.17. The upper curve refers to inclusion of the R_{pk}-layer. The scattering of measurement data is somewhat higher than in other cases.

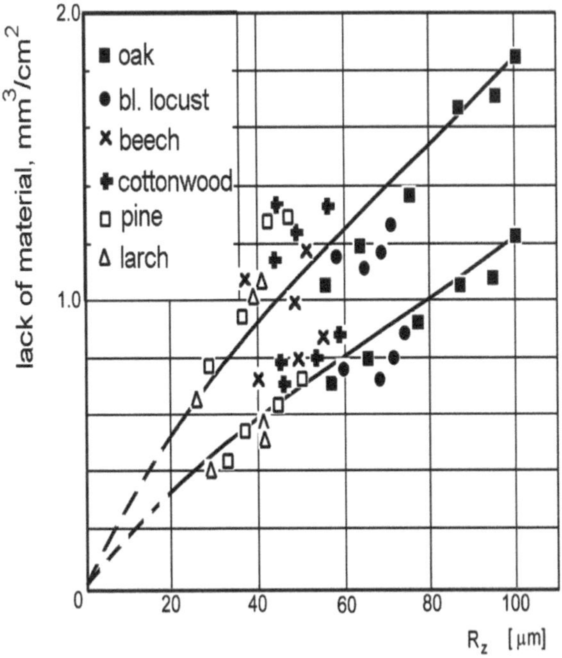

Fig. 8.17 Relationship between irregularity depth R_z and the lack of material on the surface with and without the R_{pk} layer [8]

8.6 The Use of Structure Number

The determination of the structure number ΔF for wood species has become feasible by using the data from Table 8.1 and Eq. (8.3). The structure number gives a unique characterization of a particular wood species concerning its internal structure. Furthermore, differences caused by the area where the tree grew will be taken into consideration. Therefore the structure number is expected to have a definite correlation with the roughness parameters, regardless of the wood species and the area where they were grown.

Samples were processed on a special super surfacer (so-called horizontal surfacer), where the R_Z component of roughness caused by machining did not exceed 10 μm. The relationship established by evaluating 10 different wood species is shown in Fig. 8.18 [7].

This curve demonstrates the least surface roughness that can be achieved in practice as a function of the structure number. The relationship can be described with the following empiric equation:

$$R_z = 1.22 \cdot \Delta F^{0.55} \tag{8.15}$$

To calculate the structure number requires the size and specific number of vessels and tracheids. Additional small specimens to determine the structural properties were cut from each specimen that had been used to measure roughness.

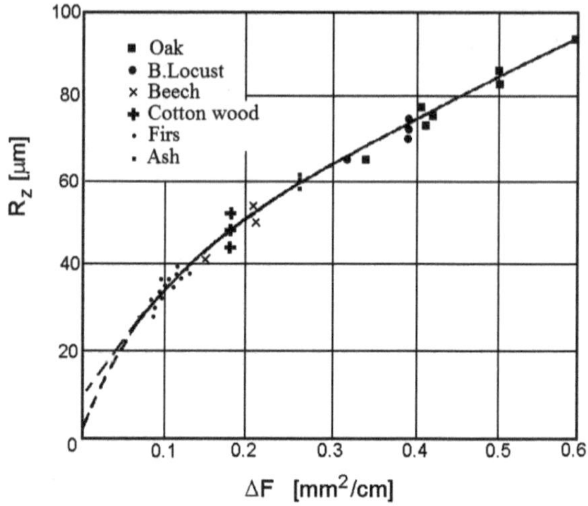

Fig. 8.18 Relationship between irregularity depth R_z and the structure number ΔF based on the parameters of 10 wood species [8]

Because the structure number is sensitive to the accuracy of experimental data, a combined image processing method and light microscope method was used. The image processing method alone did not generally give results that were accurate enough. The data are summarized in Table 8.1.

To separate the roughness components, three 20 by 5 cm samples were tangentially cut from each wood species. After machining they were finished by a special finishing machine. The finishing was repeated until the profile was flat and suitable for evaluation.

Establishing the finished surfaces, the same samples were subjected to milling operations using various cutting speeds between 10 and 50 m/s. These surfaces were evaluated with the common surface measuring methods.

A hypothetical base line was first established on the finished surfaces. The corresponding R_z'-value was calculated taking only the positive amplitudes into consideration (Fig. 8.19). This is the roughness component due to woodworking operations. Knowing the overall R_z value and subtracting the roughness component due to woodworking, we get to an R_z value due to the internal structure of wood.

Identification the machining roughness enables us to calculate and plot anatomical roughness which is the smallest roughness that can be achieved on a given sample.

Figure 8.20 shows the anatomical roughness and the actual roughness as a function of structure number at two different cutting speeds.

A general relationship for Fig. 8.20 can be expressed in the following empirical form:

$$R_z = \left(123\Delta F^{0,75} + 35e_z^{0,6}\right) \cdot \left(1 + \frac{50 - v_x}{50} \frac{0,1183}{\Delta F^{0,83}}\right) \tag{8.16}$$

$$10\ m/s \le v_x \le 50\ m/s$$

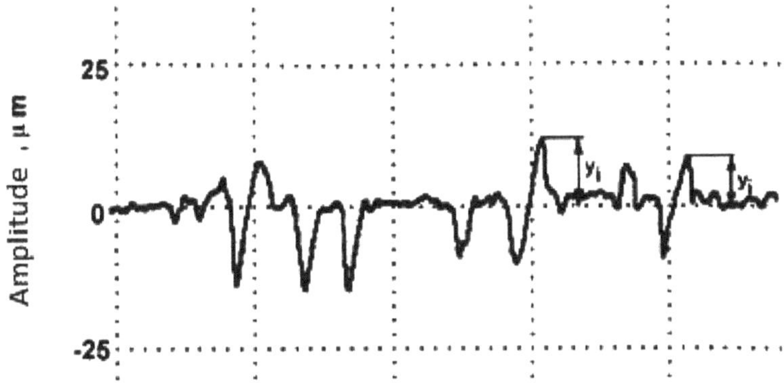

Fig. 8.19 Measurement of the roughness component due to machining [9]

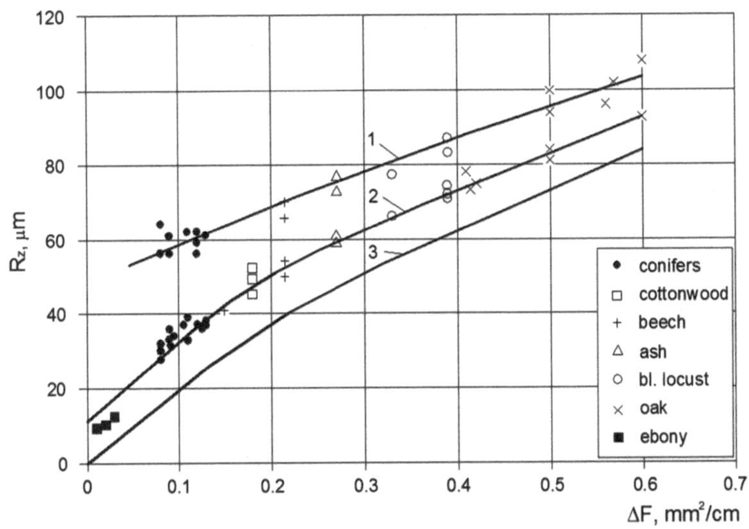

Fig. 8.20 Irregularity depth R_z, related to different cutting speeds in relation to the structure number ΔF **1**-cutting speed 10 m/s; **2**- cutting speed 50 m/s; **3**-anatomical roughness [9]

where ΔF must be substituted in mm^2/cm, e_z in mm, and v_x in m/s. The third part of Eq. (8.16) as well as the curves clearly illustrate that the softer pine wood is more sensitive to a decrease of the cutting speed. This phenomenon can be explained by the lesser local rigidity of pine.

Using a sufficiently high cutting speed, it appeared that the surface roughness will mainly be determined by the internal structure of wood.

In the following we discuss surface roughness parameter ratios which show uniquely defined correlations with the structure number ΔF.

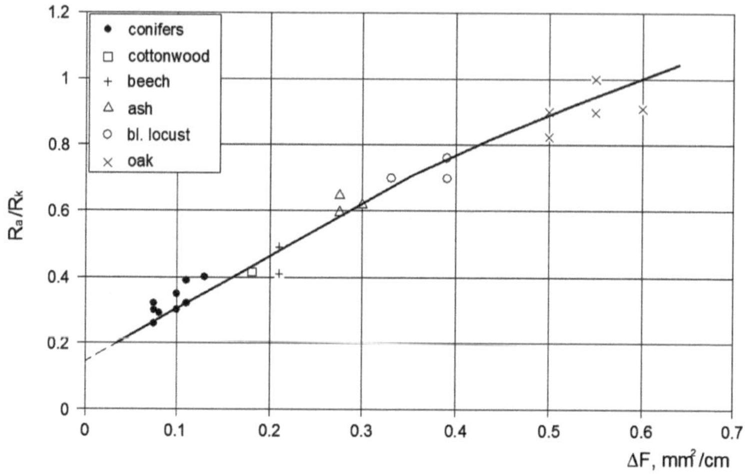

Fig. 8.21 Correlation between the relationship R_a/R_k and the structure number ΔF [9]

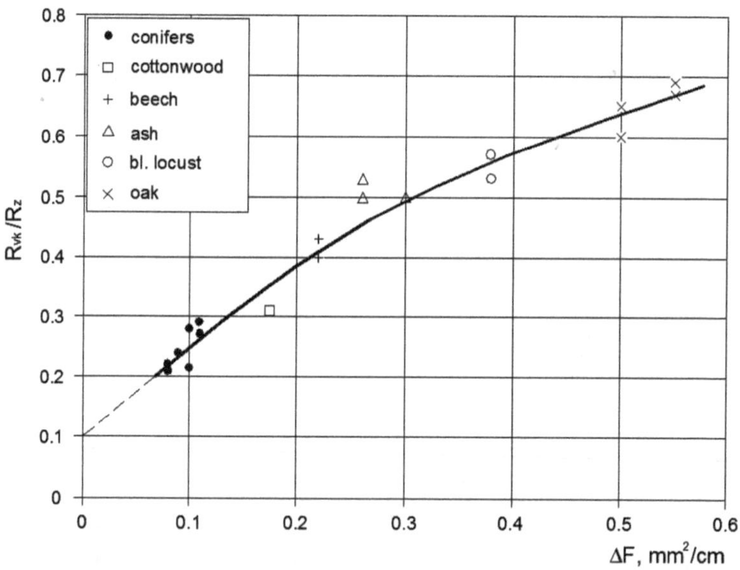

Fig. 8.22 Correlation between the relationship R_{vk}/R_z and the structure number ΔF [9]

Figure 8.21 illustrates the correlation between the relationship R_a/R_k and the structure number ΔF. The anatomical structure of wood causes a fivefold variation in the R_a/R_k ratios. This leads to the conclusion that wood species cannot be compared on the simple basis of surface roughness.

Figure 8.22 shows the correlation between the structure number ΔF and the R_{vk}/R_z ratio. Here we have a threefold increase in this ratio.

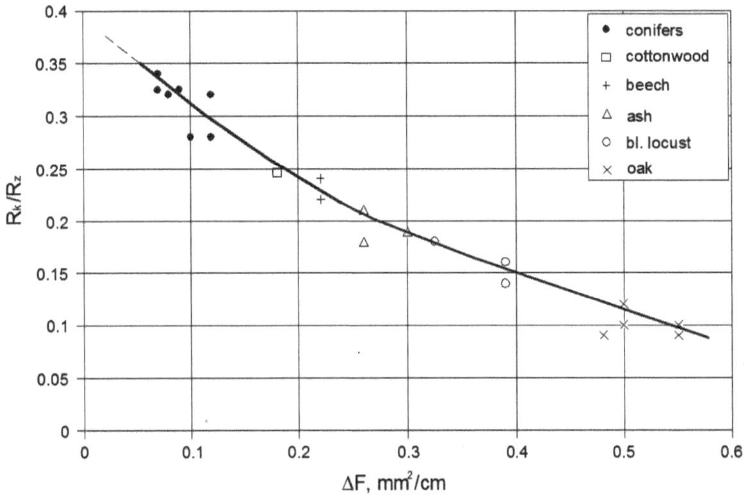

Fig. 8.23 Correlation between the relationship R_k/R_z and the structure number ΔF [9]

Finally we examined how the core depth of the material ratio curve (R_k) affects the surface roughness as a function of the structure number ΔF.

Figure 8.23 demonstrates that the value of R_k influences the roughness to a greater extent in soft wood species. It should be noted that the correlation curve in Fig. 8.23 is valid for sharp tools only. The value of R_k is dependent on tool sharpness for all wood species.

8.7 Effects of Tool Wear on the Surface Roughness

It is a well-known fact that enhanced wear of the tool increases surface roughness. This is the ultimate practical reason why tools are re-sharpened after a certain working time or cutting length.

Blunt tools with a bigger edge radius transmit bigger forces onto the material causing cell fractures. The material in front of the tool travels a longer distance going around the edge. The forces transmitted to the chip at the detachment point cause a fracture of elementary particles. Fractures beneath the cutting level are primarily expressed in the Abbott parameter R_k. Therefore, this parameter is expected to be highly sensitive to tool edge deterioration.

Figure 8.24 shows the changes in the R_k parameter of four different wood species when using two different tool edge radii.

It is clearly visible that the parameter R_k showed a doubling in each case in comparison to cutting with sharp edges. These data show that the parameter R_k gives a good feedback on the deterioration status of the tool edge.

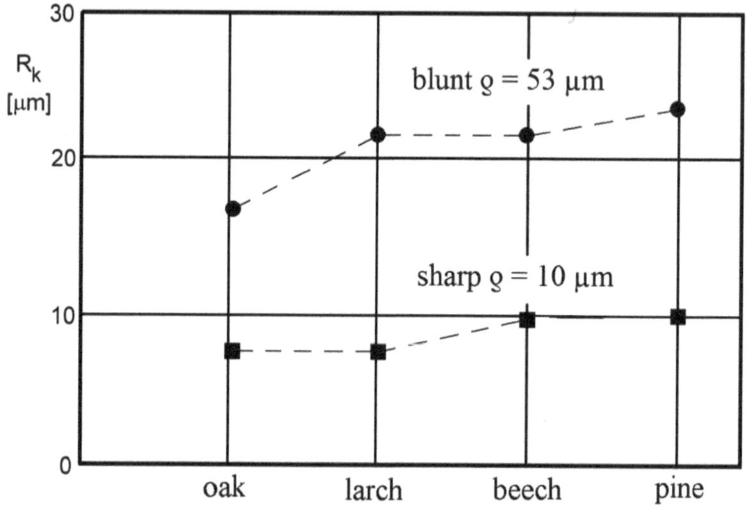

Fig. 8.24 Changes in the R_k value when using sharp and blunt tool edges in relation to four different wood species [8]

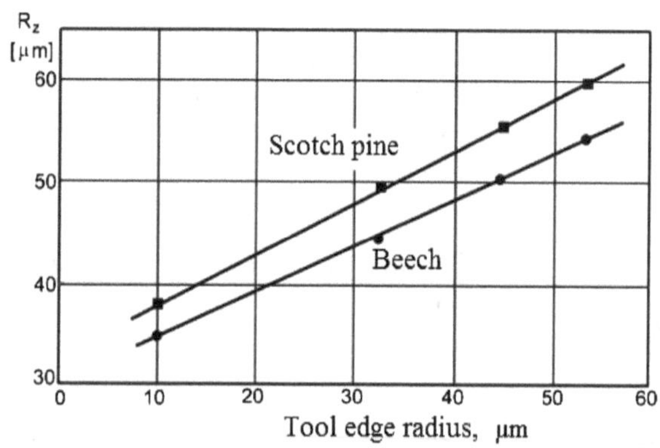

Fig. 8.25 Correlation between the parameter R_z and the tool edge radius [8]

The tool edge radius usually increases the roughness parameter R_z in a linear way. Figure 8.25 illustrates the correlation that was established by testing Scotch pine and beech samples.

We have already mentioned the compressing and destroying effect of a blunt tool edge. The compression of the surface can be observed at an approximately hundredfold microscopic magnification. This phenomenon on a sample of machined Scotch pine is shown in Fig. 8.26 [14]. At the same time, a sharp tool edge cuts a clear surface without compressed layer.

The permanent deformation reaches the same depth at all places and its depth depends on the quantity of the expelled material (depth z_0 in Fig. 8.9).

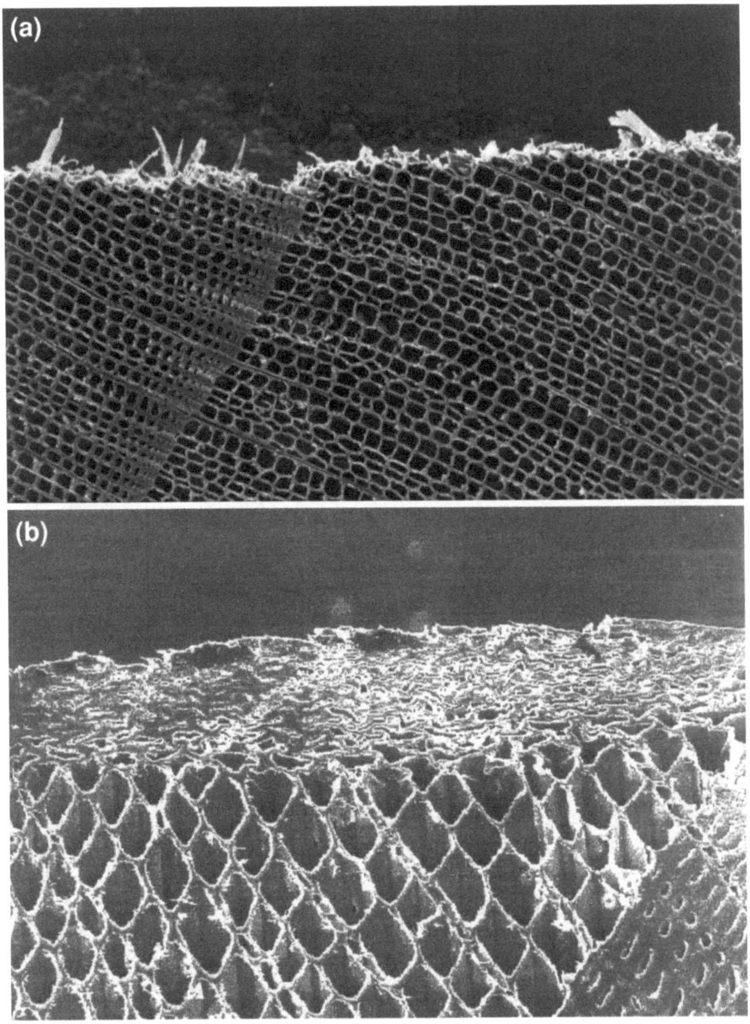

Fig. 8.26 Clear cut of a sharp tool edge (**a**) and compression of the surface as a result of a blunt tool edge (**b**) [14]

The cell walls get essentially damaged in the compressed layer therefore, this layer loses its stability in all aspect. It will have poor mechanical strength and low abrasion resistance, humidity will cause its swelling to various extents.

Furthermore, an earlier observation has shown that the compressed layer depends also on the chip thickness: using larger chip thicknesses, the same blunt edge exerts more pressure on the bottom layers and the compressed layer increases [15]. This observation further supports the advantage to use small chip thickness for surface finish. The use of oblique cutting may further decrease the damaged layer thickness.

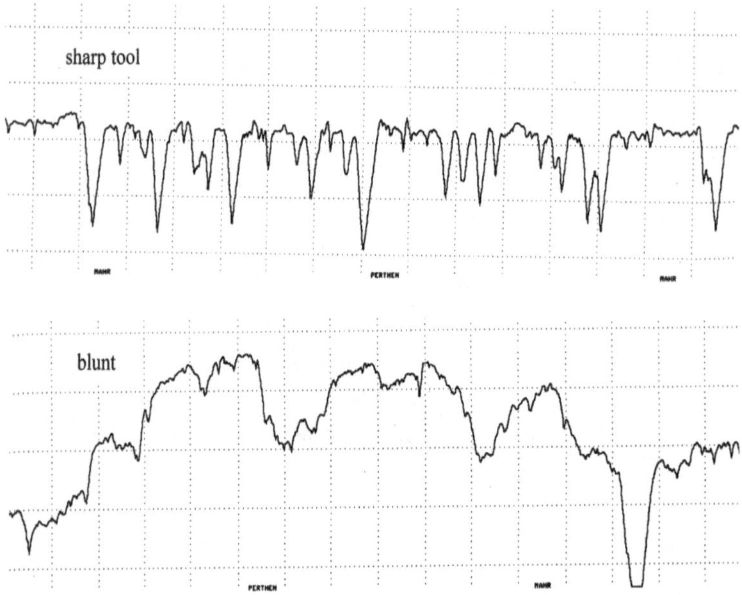

Fig. 8.27 Surface roughness profiles of oak machined with blunt and sharp-edge tools

Using a blunt tool edge for machining of large-vessel wood species this may result in surface waviness due to compression. Oak vessels can have diameters up to 250–300 μm; where the edge of a blunt tool can fit in. The edge will not only crumple but also push the material. These motions cause compression and waviness in the upper layer, the majority of the large vessels disappear from the surface due to the compression. Figure 8.27 shows surface profiles of an oak sample cut with blunt and sharp-edged tools.

When a sharp-edge tool is used, the surface is even; valleys are only caused by cut the vessels and tracheids. Using a blunt tool ($R = 53$ μm), produces an extraordinarily wavy surface, clogging the majority of the vessels. Consequently, the surface roughness alone is not always sufficient to completely characterize surface quality.

8.8 Scattering of Roughness Data

As it was established in the previous sections, main part of the resultant roughness originates from the internal structure of wood. Cavities in the wood are cut at different angles and positions during machining, which leaves valleys on the surface. The position of the surface is accidental to the position of the vessels, early and late wood. Therefore accidental effects are also present besides effects from deterministic variables which are known. This explains why roughness

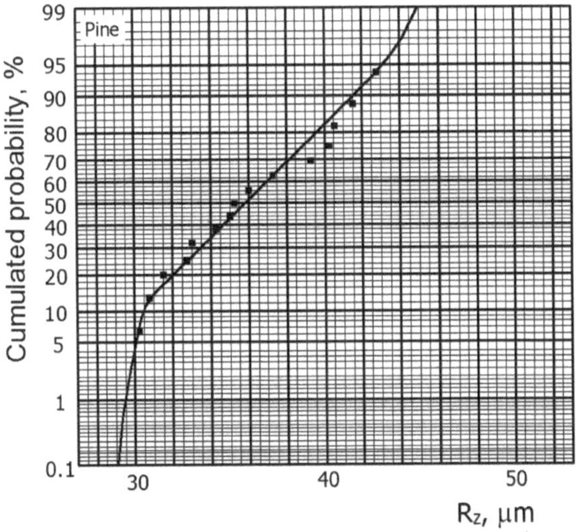

Fig. 8.28 Measurement data distribution for a Scotch pine sample [11]

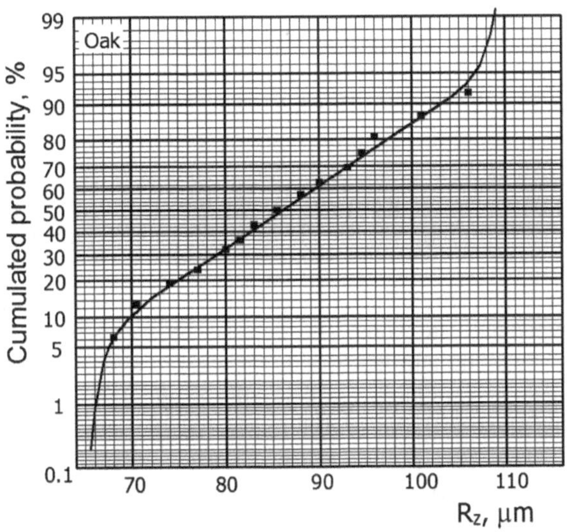

Fig. 8.29 Measurement data scattering of an oak sample [11]

parameters are always scattered around a mean value. The scattering can be determined with statistical methods using sufficient measurement data.

In order to ensure a normal data distribution, quite many, at least 50 measurements data are required for the statistical evaluation. It is practical to plot the data as a cumulated frequency curve in a probability net, since in this way straight line is achieved. Relevant plotting for Scotch pine and oak is shown in the Figs. 8.28 and 8.29.

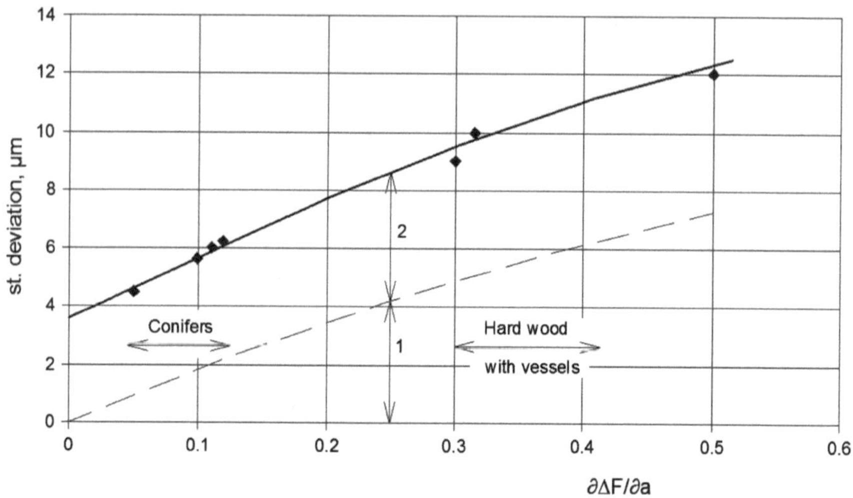

Fig. 8.30 Standard deviation of the roughness parameter R_z as a function of the characteristic number ∂ΔF/∂a for different wood species **1**—component due to structural difference in early and late wood, **2**—component due to occasional placing of cutting plane to vessels

The most part of the curves are straight with a deviation from the straight line only at their ends. This has a simple physical explanation. In practice, unlimited low and high values do not emerge as the theoretical distribution would require. Distributions like this are called uncompleted distribution

The curves clearly display the median (mean) value and the standard deviation: $R_z = 36 \pm 4.5$ µm in the case of Scotch pine and $R_z = 76 \pm 12$ µm for oak wood.

Using the curves in Fig. 8.7 we can examine the accidental effects of early wood and late wood. The slope of the curves ($\partial \Delta F/\partial a$) describes the variability of early wood and late wood, for a given species, which obviously influences their scattering. Figure 8.30 shows the standard deviation of the R_z values in relation to the slope $\partial \Delta F/\partial a$ for different wood species.

The curve is not linear but it does not significantly deviate from a straight line. The intersection ($\sigma = \pm 3.5$ µm) that belongs to the value $\partial \Delta F/\partial a = 0$ theoretically corresponds with the scattering, due to the accidental position of the surface relative to the vessels and tracheids. The question is whether this value remains constant or not for all wood species. It is very likely not to remain constant; its value will increase in species with large vessels.

The relative value of the standard deviations is worth to examine in relation to the mean R_z value. The test results for different wood species are shown in Table 8.2.

The Table shows that the relative scattering of the surface roughness of different wood species is surprisingly identical, it is dominantly in the range 0.13–0.14. This makes it very easy to estimate the standard deviation.

Table 8.2 Values of relative scattering for different wood species

Wood species	$\sigma/\overline{X}$
Scotch pine	0.13
Larch	0.14
Poplar	0.125
Beech	0.12
B. Locust	0.135
Ash	0.14
Oak	0.14

Finally, the main new recognitions and conclusions on the problems of wood surface roughness can be summarized as follows.

- The total surface roughness can be divided into two component: the first part is the component due to machining and the second part is the component due to internal (anatomical) structure.
- In the present day practice most of the roughness is originated from the roughness component due to internal structure.
- The derived structure number is based on the sizes and specific numbers of vessels and tracheids of the wood in question and further on portion of early and late wood.
- The proposed structure number shows strong correlation with the attainable surface roughness.
- Different roughness parameter ratios show definite correlation with the structure number. This finding further stresses the beneficial use of the structure number uniquely characterizing the different wood species.
- Among the different roughness parameters interrelations are found.
- The lack of material in the rough surface can be expressed as a function of surface roughness.
- The core component of the total roughness R_k is a good indicator to predict edge dullness.
- Using the structural properties of early and late wood, a characteristic number B can be defined which has a strong correlation with the expected relative deviation of the structure number ΔF.
- The standard deviation of the roughness parameter R_z is in strong correlation with the characteristic number $\partial \Delta F/\partial a$, while the relative value of the standard deviation practically remains constant.
- The standard deviation is originated from both the structural difference in the early and late wood and the occasional placing of cutting plane to vessels.
- An increasing cutting speed reduces the surface roughness, first of all by diminishing the R_{vk}-values.
- The soft wood species are more sensitive to the change of cutting velocity concerning surface roughness.

References

1. Schadoffsky, O.: Objektive Verfahren zur Beurteilung der Oberflächenqualität. Tagungsbericht Bielefeld, (1996)
2. Devantier, B.: Prüfmethode zur objektiven Bewertung der Rauhigkeit und Welligkeit von Holzwerkstoffen. Abschlussbericht IHD Dresden, (1997)
3. Westkämper, E., Hofmeister, H.W., Frank, H.: Messtechnisches Erfassen und Bewerten von Massivholzoberflächen. Abschlussbericht AiF Projekt 9681, Braunschweig (1996)
4. Csiha, Cs.: Faanyagok felületi érdességének vizsgálata „P" és „R" profilon, különös tekintettel a nagyedényes fafajokra [Surface Roughness of Wood Using „P" and „R" Profiles for Big-Vessel Hardwood Species], PhD Diss., Sopron (2003)
5. Csiha, Cs., Krisch, J.: Vessel filtration–a method for analysing wood surface roughness of large porous species. Drev. Vysk. **45**(1), 13–22 (2000)
6. Fujiwara, Y., Fujii, Y., Okumura, S.: Effect of removal of deep valleys on the evalution of machined surfaces of wood. Forest Prod. J. pp. 58–62 (2003)
7. Magoss, E., Sitkei, G.: Strukturbedingte Rauheit von mechanisch bearbeiteten Holzoberflächen. Möbeltage in Dresden, Tagungsbericht S. 231–239 (2000)
8. Magoss, E., Sitkei, G.: Fundamental relationships of wood surface roughness at milling operations. Proceedings of the 15th International Wood Machining Seminar, 437–446 (2001)
9. Magoss, E., Sitkei, G.: Optimum surface roughness of solid woods affected by internal structure and woodworking operations. Proceedings of the 16th International Wood Machining Seminar, Matsue, 366–371 (2003)
10. Magoss E., Sitkei, G., Lang, M.: Allgemeine Zusammenhänge für die Rauheit von bearbeiteten Holzoberflächen für Möbel. Möbeltage in Dresden, S. 273–279 (2004)
11. Magoss E., et al.: New approaches in the wood surface roughness evaluation. Proceedings of the 17th IWMS Rosenheim, 251–257 (2005)
12. Magoss, E.: General regularities of the wood surface roughness. Progress Report No.1, University of Sopron, p. 38 (2008)
13. Butterfield, B., Meylan, B., Peszlen, I: A fatest háromdimenziós szerkezete–three dimensional structure of wood. English–Hungarian edn. Faipari Tudományos Alapítvány. Budapest, p. 148 (1997)
14. Fischer, R., Schuster, C.: Zur Qualitätsentstehung spanend erzeugter Holzoberflächen. Mitteilung aus Institut für Holztechnik der TU Dresden (1993)
15. Санев, В.: Обработка древесины круглими пилами (Woodworking with circular saws). Изд. Лесная Пром., 1980

Index

DOI: 10.1007/978-3-642-29955-1,

FSC
www.fsc.org
MIX
Papier | Fördert
gute Waldnutzung
FSC® C083411

Kolibri 360 GmbH
Ferdinand-Jühlke-Straße 7
99095 Erfurt, Deutschland
produktsicherheit@kolibri360.de